clarisse desiles

japan
today

translated by
philip parks

64 pages of colour photographs
by jacques marthelot

16 maps and plans

éditions j.a.

summary

Cover (front):

A famous Kabuki actor about to go on stage, to portray a samurai warrior or perhaps an empress or a geisha; he is equally at home as an embodiment of masculine strength or of feminine grace, intelligence and beauty.

Cover (back):

The Golden Pavilion (15th century) was a refuge for world-weary aesthetes in search of refinement and peace.

*Previous pages:
Tokyo is a fascinating kaleidoscope of a capital: its older quarters pulsate with colourful life.*

*Following pages:
Whether clinging to the terraced hillsides or carpetting the plains, ricefields like these are a familiar feature of Japanese agriculture.*

This guide-book has been compiled with the help of Carlo Salvadé.

panorama

8	**the land and its people**
8	the isles of the rising sun
9	a violent yet a gentle place
13	Japan: rural and industrial
17	"groups" and "cadres"
20	**Japan through the centuries**
20	the jomon period
20	the yayoi period
20	chinese influence
22	kyoto becomes the capital
22	feudalism
24	the kamakura period
24	the muromachi period
25	the portuguese arrive
25	the edo period
25	the meiji era
26	japan as a world power
28	japan between the wars
28	pearl harbour, hiroshima, and the american occupation
29	the san francisco treaty
29	a renascent japan
29	japanese institutions
32	**the elusive spirit of japan**
32	japanese etiquette
35	**religion**
36	the shinto pantheon
37	confucianism
40	buddhism
41	catholics and protestants
42	**the life of mr. yamagata**
42	the child is king
44	school and university
46	mr. yamagata's professional and private life
51	**the japanese woman**
56	**the japanese home**
59	**the arts in japan**
59	"haniwa" and the great tombs
60	kyoto and "yamato"
61	kamakura: the influence of zen
61	the muromachi period
62	the momoyama period
62	the edo period
64	from the meiji era to the present day
64	the japanese garden
67	ikebana
68	the mystery of lacquerwork
68	the art of present giving
70	**literature and the theatre**
70	literature
73	the "kabuki" theatre
76	the "noh"
80	the "bunraku za" or puppet theatre
80	the japanese cinema
84	**the japanese table**
85	**the tea ceremony**

town by town site by site

88	abashiri
89	akan
92	amano hashidate
92	aso (mount)
93	beppu
96	fuji-hakone-izu (National park)
99	fukuoka
100	gifu
102	hakodate
102	himeji
104	hiroshima
106	hokkaido (island of)
109	honshu (island of)
111	ise shima
112	kagoshima
113	kamakura
117	kanazawa
117	kobe
119	kotohira
120	koya san
121	kurashiki
124	kyoto
141	kyushu (island of)
144	matsushima
145	miyajima (itsukushima)
147	miyazaki
148	nagasaki
152	nagoya
153	nara
157	naruto
157	nikko
159	noboribetsu
160	okayama
160	osaka
164	sapporo
165	sendai
166	shikoku
166	takamatsu
168	takayama
170	tokyo
192	yokahama

japanese journey

196	**getting to japan**
	where to get information?
196	air services to japan
197	when to go
197	some useful addresses
200	**planning the visit**
200	organised tours
201	travelling independently
204	japan by train
205	japan by plane
205	taxis in japan
208	six suggested itineraries
217	**everyday life in japan**
217	times and calendar
217	public services
218	good advice and good manners
220	**accommodation**
224	**japanese food**
229	**sport**
232	**a thousand souvenirs**
233	a few words of japanese
237	**index**

the land and its people

■ Modern and traditional, such is Japan today. Modern in order to retain her position as a world power; traditional in order to preserve balance, harmony and individuality—to draw strength from the past in order to face the future.

The romantic image of Old Japan, beloved of writers like Pierre Loti, has been replaced by one of brand-new modernity. People have almost ceased to associate Japan with geishas and idyllic landscapes of ricefields and wooden houses. New terms such as economy, productivity, statistics, "Japan the Third Great Power", conjure up a picture of a highly industrialised country. Thanks to the mass media this new image is fast replacing the old one. Tourists are hastening to visit a country that has changed in a hundred years from a medieval into a major world power—so completely that the American economist Herman Kahn predicts that the 21st century will belong to Japan. People are hastening to visit Kyoto and Nara, to see the ancient temples before it is too late and a land of history has become a land of science-fiction.

Yet in present day Japan only 16 per cent of the land is covered by industry—those 16 per cent of flat plain that lie mainly between Tokyo and Kitakyushu, where some 75 per cent of the population live. The rest of the country still consists of the traditional landscape of mountains, hills, terrace cultivation and ricefields. Only a few miles from the vast modern buildings wooden houses stand reflected in the waters of the paddy-fields, just as they do in engravings of Old Japan.

Right from the beginning of its civilisation Japan has assimilated foreign influences, but always without effacing the preceding ones. The Japanese do not reject, they carefully store and preserve. This is why present-day Japan is so fascinating to the visitor from abroad—life there is so extraordinarily varied, and of course some aspects are more appealing than others.

For millions and millions of Japanese life is a matter long days of hard work and a struggle for qualifications and a job. But there are compensations, calm evenings, the traditional bath, meals taken in the traditional dress, the 'yukata', and rest and relaxation in rooms with traditional decoration and furniture.

Japan is still the land of the "Kabuki" theatre and the Japanese still revel in the old world setting of the samurai films that are shown every evening on three of the eight television channels. Japanese women, who wear Western dress most of the time, delight in occasions like weddings and visits to the theatre when they wear their beloved kimono. Many firms organise courses of instruction in the tea ceremony for their young women employees.

Japan is a land where forty million people will think nothing of travelling to visit famous national beauty-spots at festival times, a land of shrines thronged with pilgrims, of peaceful monastic retreats, of gardens strewn with rocky outcrops and sculptural vegetation. The sunsets over Mount Fuji are as unforgettable as the cherry trees in blossom — beauty now complemented by the "Shinkansen", the fastest trains in the world whose lines criss-cross the country, and by gigantic ultra-modern hotels where traditionally decorated rooms delight the traveller of today...

The isles of the rising sun

...For present-day Japan has emerged out of centuries of tradition. It is a country pulsating with life, a materialist place proud of its recent achievements but yet at the same time imbued with a love of beauty and of nature, with a concern for refinement and a taste for peaceful harmony—a place where the visitor from abroad, disconcerted perhaps but fascinated too, is constantly wondering how all these contrasts of past present and future and West and Far East can co-exist so happily together.

East of China, across the sea from Korea and Manchuria, lies the Nipponian archipelago (Dai Nippon = great land of the rising sun). It consists of thousands of islands totalling some 370,000 square kilometres, about one and a half times the size of Britain, forming an arc more than 2,000 kilometres long. Formerly known as "The luxuriant reedy plain, the land of newly cut rice...", "The land of the flowering cherry," or "Omikuni" (The Great Land), the four islands of Hokkaido to the north, Honshu, Shikoku in the middle, and Kyushu to the south are bounded on the east by the Pacific Ocean, to the west by the Sea of Japan and to the north by the Sea of Okhotsk. During the tertiary period the archipelago was probably joined to the Asian land-mass by its northern and southern extremities and the Sea of Japan was simply a vast lake. Today the Straits of Tsushima (150 kilometres wide) separate the island of Kyushu from Korea; Hokkaido to the north is only some 300 kilometres from the coast of the Soviet Union. At its widest point the Sea of Japan is 900 kilometres across. The islands of Japan can be thought of as the peaks of a deeply submerged mountain range. The seas off the coasts plunge to great depths—6,000 metres near the Kuriles and 8,500 metres off the eastern coasts.

A violent yet a gentle place

A land of violence and gentleness, of extremes of heat and cold, of bare rocky landscapes and lush verdure, of islands where sea and land intermingle, Japan first captures the eye and then the heart. It compels attention, a scrutiny of its features, of the difficulties it presents. The visitor from abroad finds that he gradually becomes fascinated by the landscape,

---

FROM THE JAPANESE CONSTITUTION

■ *The Emperor is the symbol of the State and of the unity of the people, deriving his position from the will of the people, with whom resides sovereign power.*
The Japanese people forever renounce war as a sovereign right of the nation, or threat or use of force as a means of settling international disputes.
The people shall not be prevented from enjoying any of their fundamental human rights.
The Diet shall be the highest organ of state power and shall be the full law-making organ of the State.
The Diet shall consist of two houses, namely the House of Representatives and the House of Councillors. Both houses shall consist of elected members, representative of all the people.
Local self-government shall be fixed by law.
Executive power shall be vested in the Cabinet and shall be collectively responsible to the Diet.
The Imperial throne shall be dynastic... The advice and approval of the Cabinet shall be required for all acts of the Emperor in matters of State.

Promulgated 3 November 1946

The epitome of Japanese feminine elegance, this lovely lady in her kimono is a Utamaro print come to life.

the multifarious aspects of these islands, and by the caprices of nature which have been such a potent force in shaping the character of the Japanese people.

A bird's eye view of Japan is a daunting experience. It is staggering to see in such a narrow space, edged on both sides by rocky coasts or long sandy beaches, so many mountain ranges intertwined, so many volcanic craters and narrow valleys, and such a scattering of islands breaking the even levels of the Inland Sea.

There are mountains everywhere. Some three quarters of the Japanese land surface is very hilly and there are many bare craggy skylines. The highest peaks (3,000 metres) are in the "Chubu" region of the Japan Alps—an area which includes Mounts Fuji and Hida and is strongly reminiscent of European alpine landscape.

29,000 kilometres of coastline

Some 27 per cent of Japan consists of volcanoes. Of a total of 265 there are 50 active or semi-dormant ones. The last time Mount Fuji (3,776 metres: Japans highest peak) erupted was during the eighteenth century. There are frequent rumblings from Mount Asama, another peak nearby. The main volcanic areas are the southern part of Hokkaido (Mount Daisetsu), the centre of Honshu (Mounts Fuji, Mihara, Asama) and the island of Kyushu, where Mounts Aso and Kirishima emit spectacular—and acrid—sulphurous smoke from time to time.

Some 16 per cent of the land surface consists of plains as we have already noted. These either extend along the coasts or perch—a few square kilometres of flat land—up in the mountain ranges. There are many mountain streams which bring alluvial deposits down with them. They flow, often very swiftly, down steeply-sloping valleys towards the coasts where their waters are directed into irrigation channels or, where these do not exist, lose themselves in a network of small watercourses.

Following an age-old tradition these Japanese students come to seek spiritual refreshment amid the cherry blossom in the gardens of the Heian Shrine (above). It's a far cry from the back-breaking work of planting out the rice shoots.

The plains themselves are surrounded by alluvial terraced hillsides. They vary in aspect—from the wide valleys of Hokkaido to the narrow ledges that characterise the alpine regions. The most beautiful terraced areas are to be found along the coasts.

Japan has some 29,000 kilometres of coastline, an astonishingly large amount in relation to its area. The coasts of the Sea of Japan are characterised by inlets, pink rock formations and fine sandy beaches and, further north, by wide banks of sand dunes rising to considerable heights. The rocky Pacific coastline is deeply indented and great fractures have given birth to deep bays (at Sendai, Tokyo and Nagoya) and peninsulas (Izu). On the Inland Sea, where the coast is constantly being eroded away, the deposits brought down by the rivers make up for the loss. It is here on these low-lying alluvial plains that the great industrial zone of Japan has developed.

A land submitted to natural hazards

At the same time every year, in August and September, the typhoons sweep in from the Pacific, often accompanied by tidal waves and torrential rains, and ravage the south of Honshu, Shikoku and Kyushu, claiming many lives—despite the radio warnings—and destroying thousands of houses, despite the hedges and high walls that protect them.

Volcanic eruptions are often foreseen and there are few victims nowadays. But earthquakes, which still escape prediction, sometimes have catastrophic effects. In Tokyo 5,000 tremors are registered every year; there are three noticeable ones every month. The regions most affected are Tokyo and Osaka. Quakes under the sea bed produce high tides and tidal waves 30 metres high sometimes pound the coasts.

Other natural hazards include landslides, which are frequent in Hokkaido, prolonged snowfalls in the north, and torrential rainstorms during the "plum season" (June and early July) causing floods which are sometimes followed by a period of severe drought.

But the Japanese have adapted themselves to cope with these phenomena; their consequences have been lessened thanks to Japanese inventiveness and foresight—to such an extent that visiting Japan today is neither more nor less fraught with danger than a visit to any other developed country in the world.

"A cool head, warm feet and a temperate heart" (Japanese saying).

The Nipponian archipelago lies between the 31st and 45th degrees latitude. The island of Hokkaido is in the same latitudes as the south of France. Tokyo is the same latitude as Algiers and Kyushu is on roughly the same level as the south of Morocco. Yet the Isles of Japan do not enjoy a warm and temperate climate but are subjected to both maritime and continental influences. A warm current, the "Kuroshio", flowing north from between the Philippines and Taiwan, enfolds Japan from both sides—one arm follows the coastline on the Sea of Japan as far as Hokkaido, whilst the other is halted off Tokyo by a cold current flowing down from the north, the "Oyeshio." These two currents combine to moderate the influence of the nearby (cold) Asian land-mass. In winter the Siberian anti-cyclone produces cold winds which, crossing the Sea of Japan, bring snow to the archipelago. In summer sub-tropical breezes blow in from the Pacific and bringing rain. Thus Japan enjoys a very varied climate. In winter in Hokkaido the thermometer may fall to -40°C, though this happens rarely, fortunately. The island is thickly blanketted with snow and the Sea of Okhotsk, to the north, is completely frozen over. In summer the heavy humid heat makes life so unpleasant for the inhabitants of the cities along the coast of Honshu that they flee to the mountains if they possibly can.

Such a varied climate produces a rich variety of vegetation. Japan is 68 per cent forest, which cover the mountain areas particularly. Fast-growing conifers are being increasingly widely planted (30%); they provide both building materials and paper pulp. In the south there is subtropical forest—camelias, magnolias and ilex, giving way on the mountains and in the north of Honshu, to oaks, beeches, maples, birches and chestnuts. The island of Hokkaido is planted with large conifers, notably Japanese cedars, whose trunks can attain 40 to 70 metres. On Honshu they take great pride in their cherry trees, "sakura", of which there are numerous varieties. The cherry blossom season extends over two months in Japan: they flower in Kyushu in mid-March, in Kyoto towards the end of the month, in Tokyo at the beginning of April... and from the 1st to the 15th of May in Hokkaido. On Kyushu they are proud of their tropical plants, palms, orchids, mandarin oranges and bananas. Tea is grown in the southern half of Japan. Bamboo flourishes everywhere and its yellowish green is a common feature of the landscape. Only 15 per cent of the land surface is cultivable. Ricefields are planted wherever possible. The rains which fall in the "plum season" (June and early July) make the rice grow and, in the south, make it possible to grow two crops in the year.

Japan: rural and industrial

One side of the country, facing the Sea of Japan, is relatively underpopulated and rural; whilst the other, facing the Pacific and the Inland Sea, is heavily industrialised and, for some years now, extremely populous. Between these two zones rise the Japan Alps.

The former zone, where the winters are often quite cold, is where the rice is grown—in terraced fields and on the narrow plains. Only some 15 per cent of the Japanese land surface is cultivated. Rice is the most important crop. Since 1969 the annual rice harvest has been maintained at 12 million metric tons. The Government subsidises agriculture by buying rice from the farmers at a price higher than the international rate. Thus the price paid to the growers is higher than that paid by the consumers in the shops.

Only in Hokkaido will the visitor see any signs of extensive farming patterns, elsewhere the terrain makes small-scale, highly-intensive cultivation inevitable. Despite mechanisation the Japanese countryside has preserved a traditional appearance with tiny paddy-fields and meadows. The design of farm buildings varies slightly from region to region, but farms surrounded by trees and plantations, with their glazed tiled roofs reflected in the waters of the rice fields, are to be found all over Japan. The farms are grouped together to form "muras"—a local unit roughly equivalent to a village in the Western sense. Many village families can trace their origins back to feudal times; these old families stand in a position of authority over newcomers to the area.

Due to their small size Japanese farms (except in Hokkaido) do not use much large machinery. Small power-driven cultivators are used on 74 per cent of the holdings, they are highly practical for rice-growing—but the plating-out is still done by hand.

In recent years there have been changes is the Japanse diet. People are consuming more milk, meat and fruit and vegetables than they used to. The Government is therefore adjusting rice production to fluctuations in supply and demand. Paddy-fields have been turned over to wheat and soya-beans (though 90 per cent of these foods are still imported), peas, haricot beans (the Japanese are the world's largest consumers), and cabbages.

The "Tokaido Way"

In central Honshu rice, vegetables, and wheat are grown, with tea and mandarin oranges, in the hills. In the southwest they grow rice and potatoes. In northern Honshu and in Hokkaido where winters are severe,

PANORAMA 13

rice is the staple crop; wheat is alternated with sugar-beet and potatoes. The production of apples, mandarins and grapes has trebled in the past ten years, to cater for changes in consumer demand. Japan has become one of the world's leading producers of hot-house fruits and vegetables. In February Japanese markets are bright with strawberries, which are universally popular.

Japan lacks good pasturage, so grazing animals are not common; there are some 2 million dairy cattle and about the same number of beasts raised for slaughter.

Despite Government subsidies and despite a rise in the peasants' standard of living, the past twenty years have seen a flight from the land, rural depopulation or "kaso", and a drift to the towns producing excessive urban population or "kamitsu". Before the Second World War 50 to 60 per cent of the active population worked in the countryside. After the war the percentage fell to 40. The drift of people to the big cities of the industrial belt between Tokyo and Kitakyushu, became really significant in 1950, and was a response to the growing demand for labour in the factories. By 1955, the three metropolitan zones of Tokyo, Osaka and Nagoya contained 33 million people, more than a third of Japan's population. Twenty years later, the population of the three cities was nearly 54 million, the additional 20 million being the result not merely of immigration but also of a high birthrate. In 1973, economic recession reduced the exodus from the land to some extent. In 1977 it seemed that the big cities had ceased to attract new population, for in that year forty-seven rural prefectures reported that their population had grown—a reversal of a trend that had lasted fifteen years.

There are various reasons for this: there is no longer a reservoir of young people in the countryside; the standard of living of the rural population has risen whilst the towns are plagued by housing shortages and a high cost of living, as well as long journeys between home and work.

Great cities like Tokyo remain

Japan, a land of fire:
in Hokkaido and Kyushu there are
smoking mountains and valleys where
sulphurous waters bubble infernally.

The warm spring sunshine transforms the wild landscape of Hokkaido; here the thaw is getting under way, in May.

magnets for those in search of higher education and improved employment prospects, but the great mass migration from the land to the cities seems to be over.

The "inside" of Japan, facing the Pacific and the Inland Sea, was the cradle of Japanese civilisation. The intellectual and artistic heart was located in the successive capitals, Nara, Kyoto and Tokyo. Over the past twenty years this "inside" heartland has spread down as far as Hiroshima and Kitakyushu and has become the centre of Japanese industry. The region is served by the fastest rail service in the world, the bullet-trains or "shinkansen". Its route, the "Tokaido Way", serves the main cities and industrial centres, the ports (Kobe, Yokahama) and three quarters of the population of the country.

Every year the countryside is eaten into along this route by new factories and buildings. Many peasants combine work on their farm with a job in a nearby factory. Out of five million peasant households, only 12 per cent live from agriculture alone, 24 per cent have at least one member working part-time in a factory, and 63 per cent derive the greater part of their income from non-agricultural work.

The whole Japanese economy is based on the "Tokaido Way". In August 1945, Japan, defeated, humiliated and bankrupt, faced a bleak future. Together with her colonial empire she had lost her overseas investments and was faced with the problem of repatriating millions of inhabitants into an already overcrowded homeland. She had lost her merchant fleet, she had no stocks of primary products and her agriculture was in a catastrophic state.

The economic "miracle"

Twenty years later Japan had staged a miraculous recovery. She had risen from her own ashes and had become a partner of the powers who had defeated her.

How did this economic miracle take place? It was not until 1952, under the San Francisco Treaty, that Japan recovered her sovereignty and her industrial and commercial freedom. In 1960 Prime Minister Ikeda launched his plan to "double the national income in ten years."

From 1966 to 1970 Japan enjoyed the highest rate of overall economic growth in the world, over 10 per cent per annum! What made this possible?

A combination of various factors. Primary products were readily available on the world market and prices were low. Japan immediately adopted advanced industrial techniques. Her people's ingrained habit of saving gradually provided a boost to industrial and commercial investment. She had a large, well-trained, and cheap labour force. Exports were helped by a fixed exchange-rate of 360 yen to the dollar. Above all the Japanese economic miracle was made possible by the very nature of the Japanese people themselves —their capacity for hard work, their sense of working solidarity, their loyalty to their employers, their competitiveness and their respect for hierarchy.

From 1966 onwards wages and salaries rose regularly, without—until 1972—inflationary effects. The standard of living rose steadily. Average income rose from 360 dollars in 1960 to 1,300 dollars in 1971. In 1973, the year of the world oil crisis, Japan's growth rate slackened to 2 per cent per annum. In 1975 she had a deficit of 2,676 billion dollars on her balance of payments.

The situation rapidly improved and in 1976 she had a payments surplus of 2,426 million dollars (exports: 67.2 billion dollars, imports: 64.8 billion dollars).

When faced with inflation, brought on by the oil crisis, the Japanese Government adopted a vigorous policy, ensuring Japan's technological autonomy and choosing to protect the environment (anti-pollution measures for example) rather than increase productivity, while at the same time attempting to make the country self-sufficient in food production.

On 21st October 1977 the newspapers reported "a new record for Japanese exports." During the first quarter of the 1977 fiscal year an all-time export record of 43 billion dollars was reached, a rise of 10.78 per cent compared with the previous record of 38.8 billion dollars, for the second quarter of 1976 and of 19.4 per cent compared with the first quarter of 1976. This rise was also partly due to the increased value of the yen compared with the dollar—a factor that accounted for a 9 per cent rise in the value of Japanese exports during this period.

Japan's top exports in 1977 included cars (a rise of 27.9% compared to 1976), shipping (a rise of 26.7%), photographic material and motorcycles. Her balance of payments surplus provoked strongly critical reactions from the United States and the European Economic Community. However, even if Japanese exports upset foreign markets to some extent, they do have certain beneficial side-effects. Japanese exports of industrial installations, for example, create employment in the countries that buy them. As far as Japan's controversial shipping exports are concerned, it is worth noting that Japanese orders (amounting to 80 per cent of world demand) are so high because of the ultra-competitive prices her builders are able to quote—some 20 to 40 per cent lower than European prices. Finally, it is worth bearing in mind that the overall growth of E.E.C. exports is in fact higher than that of Japanese sales abroad...

"Groups" and "cadres"

It is forecast that by 1985 Japan will achieve some 94 billion dollars of overseas investments. Will the 21st century belong to the Japanese?

The idea of the "group" is extremely important in Japanese society. A group is composed of individuals who have common characteristics, either inherited or acquired (academic qualifications, social position). A "cadre" is made up of individuals with different characteristics but in a common situation. Thus an engineer's profession is one of his characteristics. If he works as an engineer at Mitsubishi this gives him a "cadre". The "cadre" (office, company, university) is very important for all Japanese. In order to preserve the unity of the "cadre" a group consciousness must be developed vis-a-vis the world outside and one's own group must be isolated from other groups.

All problems are solved within the various "cadres"; they constitute the great strength of Japanese commercial and industrial firms.

Trade unions are formed on the basis of the individual company and are organised vertically, including members with differing qualifications and occupations. Except among teachers and the employees of the private railway companies there are no horizontal links between workers in different firms; all conflicts between employers and employees are "internal affairs." An individual enters a "cadre" and remains a member of it for the rest of his life. The "cadre" concerns itself with the family life of its members, their housing, marriages, and spare-time activities, as well as the whole range of employer-employee relationships. This gives rise to a complex hierarchy within the firm. Employees with similar qualification are nevertheless graded according to age, length of service and so forth. One of the advantages of the "cadre" system is that decisions are never taken by an individual alone but by the whole group. Thus responsibility is shared collectively, it is never an individual matter.

The disadvantages of this group system lie in the difficulty of making a rapid change in its leadership, no single individual being prepared to supplant his leader; moreover the power of the leader is restricted, since he can exercise it only through a series of subordinates.

Until quite recent years this scrupulous concern to preserve hierarchies often led to leadership being confined to the old. But a new generation of young technocrats seems to be beginning to change this... □

Hokkaido is a truly Nordic land;
in winter a place of ski-slopes
and wide horizons,
in summer hot springs gush up around its lakes.

japan through the centuries

The Jomon period
3000-300 B.C.

Little is known about the origins of the Japanese. Finds of pre-historic pottery in the Tokyo area have indicated the presence there of hunters and fishers during the neolithic period. According to legend, the first Emperor, Jimmu Tenno, great-grandson of the Sun Goddess Amaterasu (see chapter: "Religion"), lived around 660 B.C. Towards the 4th century the Ainu, white-skinned hairy people, probably from Western Siberia, are said to have landed on the coasts of Honshu; they probably settled in the north of the island but were gradually pushed out, over the centuries, into Hokkaido, where their descendants still survive among the local population (see entries for "Hokkaido" and "Noboribetsu").

The Yayoi period
3rd century B.C.

The Japanese islands were first settled during the 3rd century. Opinions differ as to the origins of these early settlers. Some observers argue from the proximity of Korea and the relative ease of access that the Japanese are of Mongol descent. This view is supported by the physical similarities of the Japanese to the Mongol racial type—breadth of skull, skin colour, lack of beard and slanting eyes. Moreover, pieces of jewellery and cooking implements have been found that are probably of Korean origin. Others support the theory that the Japanese are descended from the Malays. The Malays were great voyagers and could have been brought up to these islands by the "kuroshio" current. Indeed many Japanese are dark-skinned like the Malays. It seems likely in fact that both peoples landed in Japan, settled there and inter-bred. It is now thought that other races also came and mingled with the settlers to produce the present-day Japanese people. These immigrants are said to have brought their knowledge of wooden agricultural tools and of the cultivation of rice. During the 1st century B.C. there was another wave of Mongol invaders, who brought a knowledge of metals, spears and mirrors of bronze.

The centres of settlement were the Izumo region, by the Sea of Japan, the island of Kyushu, and Yamato (near Nara) which soon became the principal state. The accounts that have come down from these distant times often contain a mixture of historical and mythical figures—such as that of the legendary Queen of Kyushu, Jingu Kogo, who is said to have conquered Korea during the 3rd century A.D. At first there was no fixed capital, for according to Shinto belief death is a kind of defilement and the sovereign had to abandon the palace of his predecessor and build another elsewhere.

Chinese influence

During the splendid T'ang dynasty there were many immigrants from China and the innovations they introduced were enthusiastically adopted by the Japanese. It was during the 6th century that Buddhism brought the culture of China to Japan, with its sacred texts, its works of art and its temple architecture. Devout Japanese converts journeyed to China to study their new religion. In order to equal the Chinese Emperor, known as the "Emperor of the West", during the 7th century the Emperor of the State of Yamoto took the title of "Sovereign of the Eastern Empire" (of the Rising Sun: Hi no moto). Crown Prince Shyotoku sent parties of young scholars to China to study and absorb its culture. In 710 the Court broke with tradition and took up permanent residence at Nara, where a city was built on the model of Chang'an, the Chinese T'ang capital. At this time power was slipping from the hands of the Emperors into those of the Regent Prime Ministers. This latter office was filled for a long time by members of the Fujiwara family.

CHRONOLOGY

JAPAN	WEST
3,000 B.C. Jomon period; clay figurines.	Neolithic period; pottery, lake villages.
1,000-500 B.C. Legendary foundation of the Empire by Jimmu Tenno 600 B.C.	Etruscan civilisation—legendary foundation of Rome (753)—Greek civilisation 6th-4th cents. B.C.
300 B.C. Yayoi period; beginnings of agriculture; Haniwa figurines.	500-300 B.C. Socrates (489)—Alexander the Great—Rome dominant in Italy.
100 B.C. Building of the first temple at Ise.	Julius Caesar
0-300 A.D. Yamato period—Chinese civilisation enters Japan through Korea; introduction of Chinese script.	First Hun invasion (390)—fall of the Western Roman Empire (476).
500-600 A.D. Spread of Buddhism—building of the first temples (Horyu ji at Nara, the Kakyamuni Triad, the oldest Buddhist sculpture)—Influence of Chinese culture—first Embassies to China.	Merovingian domination of the Franks.
700-800 Nara becomes the capital (710-784)—First literary works —capital transferred to Kyoto (784) — Kyoto period (794-1195)—Fujiwara family gain power (866-1160).	Arab conquest of Spain (711)—defeat of the Arabs at Poitiers (732)—Charlemagne (800).
900-1000 Fujiwara Regency Feudal period—Golden Age of Literature (Genji Monogatari).	The Holy Roman Empire (962). First Crusade.
1110 Struggle between the Taira and Minamoto families. Minamoto Yoritomo becomes first Shogun —Kamakura period (1185-1333) —Zen introduced into Japan.	Concordat of Worms (1122)—Second and Third Crusades.
1300-1400 Muromashi period.	Hundred Years' War—Discovery of America (1492).
1500 Arrival of the first Portuguese — Dictatorship of Toyotomo Hideyoshi—Persecution of the Christians (1597).	Voyage of Magellan—Formation of the Portuguese Empire—Spanish occupation of the Philippines (1565).
1600-1868 Edo or Tokugawa period (1602-1867). Meiji Restoration (1868).	Louis XIII and Louis XIV — French Revolution (1789). Queen Victoria (1837-1901).

Kyoto becomes the capital

At the end of the 8th century the Court, to escape Buddhist influence, moved to Heian, as Kyoto was then known. It remained there until 1868, although effective power resided elsewhere from the end of the 12th century. This was a period of luxury and splendour. Art and literature flourished; magnificent temples and palaces were built, and monasteries where the emperors could find refuge from the exercise of power. It was a time of civilised refinement when the aim of the nobility seems to have been to spend their lives writing poetry and making music. It was also a time when women had great power in the court. During the 9th century, links with China were relaxed and Japan gradually achieved cultural independence. The Chinese script, in which literary, geographical and philosophical texts were written, was gradually modified to suit the Japanese language. This was the period of literary journals of the famous "Tales" (see "literature"). After having for so long followed Chinese models Japanese society was finding its own individual nature.

Feudalism

Throughout this period (9th to 12th cent) the Fujiwara, who were the real holders of power, attempted to maintain a centralised system. There were two kinds of nobility —those at court in Kyoto and those who lived in the provinces. Gradually the other islands and the provinces beyond the mountains became detached from the capital. Local principalities were formed. Local rulers who had defeated the Ainu or rival landowners considered themselves sovereign, "daimyo", in their fiefs. The weaker among them sought the protection of the stronger, of whom they became vassals, "samurai". Peasants in their turn sought the protection of these "samurai". Hereditary links grew up between "daimyo" and "samurai".

Two aspects of Shinto architecture, the grace and lightness of the pagodas contrasts with the heavier sobriety of the shrines themselves.

Buddhism came to Japan during the 6th century, inspiring harmonious temples, sculpture and monasteries in secluded country settings (the Manzen ji Temple, Kyoto).

Japan was divided into small hierarchical states which, over the centuries, fought against each other and against the central power. A warrior class developed, energetic and courageous and imbued with a sense of disciplined obedience to its leaders. The education of the "samurai" was entirely a military one. If they failed in what they undertook they were dishonoured; if they were subjected to an insult that they could not avenge they stabbed themselves, committed "seppuku" or "harakiri".

In order to subdue these provincial lords the Fujiwaras called upon two noble families, the Tairas and the Minamotos who were themselves bitter rivals. In 1156 and 1160, Taira Kiyomori, having defeated his Minamoto enemies, became the most powerful lord in Japan; he married his daughter to the Emperor, "dethroned" the reigning Fujiwara and became Prime Minister himself. But in 1185 he was overthrown by the Minamotos. This epic struggle is one of the most colourful chapters in Japanese history. Even today during the "Boys' Festival', at the beginning of May, wooden and bronze replicas are sold of the helmet of Yohitsuna, one of the heroes of the struggle. The "Heike monogatari", an account of the battles of the two families was very popular during the 13th century: as the "Tale of Heike" it is now available to Western readers.

The Kamakura period

Minamoto Yoritomo, the "seiitai-shogun" or "General in Chief against the Barbarians", left the Emperor and his Court at Kyoto and, assisted by a regent "shikken", set up a campaign headquarters or "bakufu" at Kamakura in Kanto. He was the first of a long line of shoguns who were the effective governors of Japan until the Meiji Restoration of 1868. The Kamakura period saw the development of the military caste; the classic image of the "samurai" dates from this time. Buddhism, until then the religion of the aristocracy, spread to all classes of society. Chinese "Zen" however, remained confined to the military caste (see: religion).

In 1274 the Mongols, who at that time controlled Central Asia, China, and Korea, sent an embassy to the Court at Kyoto demanding a Japanese surrender. The "shogun" refused. Mongol forces landed on Kyushu, at Hakata (Fukuoka), but bad weather obliged them to set sail again. In 1281 Kubilai Khan set out again with an armoured fleet carrying 150,000 men. Just when victory seemed to be within the Mongols' grasp a "divine wind" or "kamikaze" arose and their fleet was destroyed. This success left the discontented Japanese warrior caste unmoved; they had become weakened by perpetual internal battles and impoverished by the division of their lands over the generations.

The Muromachi period

In 1331, the Emperor Go Daigo attempted to regain power (held at that time by the Hojo family) with the aid of Ashikaga Takauji and his warriors. He succeeded, but Ashikaga's followers felt that they had been inadequately rewarded and united to depose the Emperor. He in turn fled to Yoshino in Yamato, where he set up the Southern Court. Ashikaga established a new Emperor who set up the Northern Court, at Kyoto. Ashikaga took the title of "shogun" (which was to remain in his family for two centuries) and installed himself at Kyoto, in the Muromachi quarter, in order to keep watch over the Imperial Court. Despite their proximity to the Emperors the Ashikaga gradually ceased to be able to check or subdue the numerous powerful "daimyo" who owed their prosperity to families long established at Court.

The Court itself gradually grew poorer whilst the economic and cultural life of Japan prospered. The Ashikaga were great patrons of the arts. The wealthy "Zen" monasteries influenced new art forms: the "Chanoyu" (tea ceremony), landscape gardening, the "Noh" theatre.

Japan engaged in a flourishing trade with China and other Asian countries, importing silks, books and paintings, and exporting pearls and swords. The "shoguns", the monasteries and the rich "daimyo" protected this traffic. A new commercial class developed.

The Portuguese arrive

In 1543 the first Portuguese traders arrived in Kyushu; in 1549 the Jesuit missionary St. Francis Xavier followed them. By 1580 it is estimated that some 150,000 Japanese had been converted to Christianity. Together with the Europeans, Western civilisation was beginning to make inroads into Japan. Two great figures stand out at this time: Oda Nobunaga, "daimyo" of Nagoya, who seized power in Kyoto, exiled the last Ashikaga "shogun", fought the feudal lords and the Buddhist monks, and was assassinated in 1582, and Toyotomo Hideyoshi, who took power and made himself dictator. He built a great fortress in his capital, Osaka. He organised a regular troops, armed them with cannon and subdued the Satsuma clan in Kyushu and then took control of the north of Honshu. He planned a conquest of China, invaded Korea, and died in 1598. Some writers refer to him as "the Japanese Napoleon".

The Edo period

Tokugawa Iyeyasu seized power; on entering Osaka he obtained the "shogunate" from the Emperor and established himself at Edo (Tokyo), which became the "shogun" capital, the Emperor remaining at Kyoto. Iyeyasu encouraged Confucianism but became the implacable enemy of the Western missionaries. After the establishment of the Dutch trading post at Dejima (see under "Nagasaki"), he felt he could now trade with Europe without having to deal with the Christians, he now saw them merely as fomentors of trouble and chased them en masse out of Kyushu.

His successor, Iyemitsu, persecuted Christians systematically. He purged Japan of foreigners; the only ones to remain were the Dutch traders and a few Chinese and Koreans at Nagasaki. The Japanese themselves were forbidden to build ships big enough for ocean voyages, forbidden to leave Japan or to learn foreign languages. Japan entered on a period of internal peace.

Society was strictly compartmentalised. At the top were the landlords, the "daimyo", masters in their own domains but under the control of the "shogun"; they and their families had to spend six months of the year in Edo. Next came the warrior "samurai", below them the peasants, comprising 85% of the population, them came the artisans, 2%. At the bottom of the scale was the minute commercial class, despised yet envied for its wealth. Finally there were the "parias", social lepers such as butchers, slaughterers and tanners.

In Edo the rich merchants gradually extended their influence. Intellectual and artistic life became more democratic. This new bourgeoisie attended the "kabuki" and "bunraku" (see "theatre"). "Geisha" quarters grew up in the big cities. The merchants built themselves luxurious houses and surrounded themselves with precious things (ceramics and lacquerware for use in the tea ceremony).

The Meiji era

On 8th July 1853 an American fleet, under the command of Commodore Perry, dropped anchor in Edo Bay and demanded that Japan open her doors to American traders. The fleet returned the following year and succeeded in obtaining permission from the "shogun" for a few Americans to begin trading in certain ports. The British, Russians and Dutch asked for the same facilities. Japanese opinion was divided—some

favoured a policy of opening the country, whilst others called the "shogun" a traitor and claimed that he was turning Japan over to the barbarians.

In 1867 the last "shogun" abdicated and power was resumed by the Emperor Mutsuhito, who took the name of Meiji, "enlightened government". The "daimyo" made gifts of their lands to the Emperor and received indemnities and titles in return. It proved more difficult to deal with the "samurai". They gave up their right to bear arms and received money for doing so; some went into the army, others into trade. The peasants, still more or less serfs, were given their freedom and acquired ownership of the land they had worked.

Japan decided to westernise herself, to study the European nations and take from them what she judged best. Thus it is said Japan acquired a French legal system, a navy modelled on the British, an army and medical system modelled on the German, and American commercial practice. A civil service, banks, law courts and a police system were either created or drastically modernised. Compulsory schooling was introduced. In 1889 a Constitution was promulgated. The State financed the creation of an economic infrastructure: railway and telegraphic systems were begun. Commercial and industrial enterprise was subsidised; peasants who left the countryside to work in the new factories formed an important labour force.

Japan as a world power

In 1894 Japan, who had been observing the rivalry between Western powers in the East, decided to play a role there herself. A Japanese fleet destroyed the Chinese naval forces and attacked the coasts of China, gaining Formosa as a result. Korea became independent. The Russians, who frequently involved themselves in Korean affairs, looked like a threat to Japan. The Japanese attacked, by sea at Tsushima and on

*Ancient Edo lives on in Aksakusa,
the old pleasure quarter.
A throng of colourful shops,
restaurants and craft workshops
still surround the temple of Kannon, Goddess of Mercy.*

land at Mukden; the Russians capitulated in 1905 and ceded the southern half of Sakhalin to Japan. In 1910 the Japanese annexed Korea and proceeded to colonise it, a valuable acquisition with its rice and mineral wealth.

During the First World War Japan took advantage of the Europeans' neglect of the Asian market to enter it herself. She also seized German islands in the Pacific and the German trading post of Gingdao in China. In 1919, only fifty years after her emergence from the "Middle Ages" Japan took her place with the victorious allies at Versailles.

Japan between the wars

During the First World War a new ruling class of businessmen, military and intellectuals had taken over. The regime remained democratic until 1930. A new middle class, largely composed of teachers and engineers, all university graduates, gradually emerged and kept in touch with Western political ideas. Left-wing newspapers were launched; a nascent Communist party was stifled at birth; there were occasional strikes and trade unions appeared upon the scene.

The towns and cities became westernised, men adopted western dress; European influence could be seen in new buildings (e.g. the new Imperial Akasaka Palace).

In the countryside people clung to the old ways and were supported in this by the army, composed of the sons of officers, and the landed proprietors, brought up according to traditional beliefs and values. The younger officers, disgusted by the ostentatious luxury of the cities, felt repelled by the new capitalist class; they inveighed against this luxury and its Western origins. This growth of militarism coincided with Fascism and Nazism in Europe. Along with it there developed a new cult of the Emperor, "Official Shintoism".

Public opinion favoured colonial expansion in order to provide new primary and consumer goods for a rapidly rising population. In 1931 a Japanese fleet arrived at Shanghai and took possession of the city. In 1932 the puppet state of Manchukuo was set up after the conquest of Manchuria. At the same time military dominated "national governments" were replacing "party governments" at home, in the wake of a series of political assassinations.

Recent American legislation, restricting Japanese immigration, was felt to be insulting and Japan decided to drive the white man out of Asia. The anti-capitalist military came to terms with the "zaibatsu" who controlled the country's economy and without whose support military and naval expansion would have been impossible. The "zaibatsu" saw the coming conflicts as an opportunity to gain in both power and wealth.

The war with China

In July 1937 Japanese forces from Shanghai finally entered Peking, to await the capitulation of General Chiang Kai-shek. The violent invaders met with fierce resistance from the Chinese population. Japan concluded alliances with Germany in 1936 and with Italy in 1937 and soon took advantage of the successes of her new allies. The fall of France in 1940 gave Japan a free hand in Indo-China, which she invaded soon after. The United States and Britain reacted to the danger presented by Japanese expansion and launched a war of attrition by seizing Japanese tanker convoys. Japan did not have to wait much longer for China to surrender.

Pearl Harbour, Hiroshima and the American occupation

On 7 December 1941 the Japanese attacked the American Fleet at Pearl Harbour. In 1942 the Japanese advance was stopped at Guadacanal by the Americans, who themselves took the offensive in 1943. The subsequent struggle was extremely exhausting to Japan; her supply lines were gradually cut by the Americans. Despite the collapse of Germany Japan would not admit defeat. However, the two atomic attacks, on Hiroshima and Nagasaki, on 6 and 9 August 1945, finally induced her to surrender. The armistice was signed on 2 September.

For the first time in her history Japan suffered the humiliation of defeat and occupation by foreign troops. The Occupation, under the command of General MacArthur, lasted six and a half years and proved relatively mild. The Japanese army and navy were abolished. Thousands of men who had played a leading role during the hostilities were excluded from public and business life.

In 1947 a new Constitution was proclaimed. Official Shintoism was repudiated and the family structure was made more democratic. The leaders of the big corporations, held largely responsible for the war, were dismissed and their assets seized. Labour laws were introduced. A rural reform provided for a more just redistribution of land.

The San Francisco Treaty

In 1950 the United States, involved in war in Korea, turned to Japan to supply her with food and other supplies. On 8 September 1951, 47 nations (not including China or the Soviet Union) finally concluded peace with Japan by the San Francisco Treaty. Japan forfeited her colonial empire. Russia occupied the Kuril Islands and the south of Sakhalin, while the United States took Okinawa (returned to Japan in 1971) and the Ogasawara Islands.

A renascent Japan

Japan abandoned any dreams of military might and decided to become a major economic power (see "the economy"). She became a member of the United Nations Organisation in 1956, and the next ten years were marked by economic growth and an increasingly active role in international affairs. Her economic progress resulted in membership of the O.E.C.D. in 1964.

Two important events drew Japan to the attention of the whole world: the Olympic Games were held in Tokyo in 1964; in 1970 the World Fair at Osaka was an event of international importance.

Japanese institutions

The new Constitution which was promulgated on 3 November 1946, came into force on 3 May 1947 (see special entry).

The Imperial family.

Emperor Hirohito was born in Tokyo, in 1901. He married Princess Kuni, now the Empress, in 1924. He succeeded to the throne in 1926. The Emperor's hobby is marine biology. Crown Prince Akihito (b. 1933) and Princess Michiko have two sons and a daughter. The Crown Prince has frequently travelled abroad. Other members of the Imperial Family include Prince Hitachi, younger son of the Emperor, and Princes Takamatsu and Mikasa, the Emperor's brothers. The Imperial Family receives an allowance of 547,000 dollars a year from the State.

The Legislature.

The Diet is the highest authority in the State and its only legislative body. It consists of the House of Representatives (elected for 4 years) and the House of Councillors (elected for 6 years). Both houses are elected by universal suffrage (the voting age is 20); Councillors are elected according to a system combining regional and national representation.

There are five main political parties.

The Liberal Democratic Party, in power at the moment, regards as its fundamental mission "the protection of liberty, the rights of man, of democracy and of parliamentarianism." It aims to raise the standard of

The Shinjuk District, in Tokyo, is particularly lively at night, with cabarets and gaming houses and many little restaurants.

living, improve international relations, and to produce stability in economic and social life.

The Socialist Party regards itself as a mass party, the party of the working class, of those engaged in agriculture and fishing and in small and medium business. It is also supported by intellectuals.

The Communist Party fights against "American imperialist domination, and Japanese monopoly capitalism, for the liberation of the working class."

The Komeito Party was founded in 1964 to achieve the political aims of the "sokagakkai" sect (see "religion"). It states that it is a party of "prudent reform" and that it aims to establish society according to "a new conception of humanity".

The Democratic Socialist Party is opposed to all extremist ideology.

There are also a number of very small parties with very few seats in the Diet.

Executive authority is vested in the Cabinet, which consists of the Prime Minister and some twelve other ministers and is responsible to the Diet. A State Audit Office controls government expenditure. Japan is divided into forty-seven Prefectures governed by mayors who are directly elected.

The judicial power resides in a Supreme Court, eight High Courts, and a District Court in each Prefecture. □

*Springtime in Tokyo's Ueno District.
A pedestrian street crowded
with Tokyoites and visitors from the country;
the stalls are a great attraction,
as well as the big stores
and the cinemas and theatres.*

the elusive spirit of japan

■ "When you realise that after four or five years you don't understand the Japanese at all, then you are in fact beginning to get to know them"—wrote Lafcadio Hearn in 1910.

Japanese etiquette

Even today it must be recognised that the Japanese remain something of an enigma to Westerners. The Japanese rarely behave spontaneously, least of all in Japan itself. Yet a careful observer can gradually—and it often takes a very long while—build up a picture and an understanding of them. In order to grasp something of the Japanese temperament it really is essential to try to put aside Western notions of logic and analytical thinking. Ideally one should steep oneself in the myriad rules and customs that regulate Japanese society and the behaviour of its people—all very different from our own.

The Japanese give an impression of being reserved, even introverted. After all Japan is a remote country and it is only during the past hundred years that she has opened herself up to the West. For centuries, apart from brief contacts with China and the hesitant opening of a window to the West during the 18th century at Nagasaki, Japan lived entirely turned in on herself. The Japanese got used to living in their closed world. A family, even a village, develops a kind of intuitive understanding between its members; it takes a foreign presence to make it articulate.

Until 1868 Japan knew few foreigners and had little stimulus from outside. But then feudal Japan busily set about trying to "catch up" and learn from the West. Yet despite this initial attraction the Japanese quickly developed a system of self-defence; they carefully developed their social structures as a protective barrier against the seductions of the "Western barbarians."

These social structures born of Japanese Confucianism—vertical relationships within the family and work groups such as schools and offices—persist in present-day Japan; they do not encourage individual self-assertiveness. There are two common and revealing Japanese words: "nakama" and "yosomo." "Nakama" is used to describe what a group may have in common. "Yosomo" means those things that are foreign to the group. The various problems of "nakama" are always discussed by the same people; no foreign idea or opinion, "yosomo", may enter these discussions for this would direct attention outwards. Thus when a Japanese leaves his group, the protective nest of "nakama", he tends to be extremely shy and reserved.

The "geisha" is one way of dealing with the possibly unfortunate consequences of this in business relations. The geisha acts as a "hostess" at business lunches and dinners; her job is to provide a relaxed atmosphere and so make relations easier.

When confronted with a Westerner this timidity of the Japanese becomes almost impassivity. This is in fact only a way of masking his surprise and bafflement; rather than risk making a mistake he may prefer to remain silent...

The permanent search of harmony

"Loss of face" is a key notion in understanding Japanese behaviour. "Face" in an aspect of an individual's attitude, the behaviour dictated to him by his family and social position. If he doesn't act in the way his birth and situation demand then he "loses face".

In Japan there are many things which are and are not "done". A Japanese has an extremely close dependence on his family and work group. For a student to fail his examinations when his family wanted and expected him to succeed involves automatic exclusion from the family group; he has "lost face". In such a case, according to ancient Confucian precepts, suicide is quite comprehensible.

A Japanese hesitates before committing himself. Life in society is a succession of committments. When he performs a service a Japanese constantly fears that he will not do it sufficiently well and in the way required... If it's a matter of present, then to give a present of equal value solves the problem; but human relations are more complicated. Under the constant impression that they are not doing what is expected of them the Japanese seem to have a feeling of perpetual indebtedness.

What are the basic causes of all this?

By behaving always according to norms dictated by his social position the individual Japanese achieves a kind of harmony and peace. This notion of harmony is religious in origin. The achievement of harmony is very important to the Japanese; it is a prime factor in their emotional stability. In his life of constant tension, as he works unremittingly to achieve a position and to maintain it, a Japanese finds this inner balance quite essential. His life-style is marked out by a series of constraints —anxiety about what people will say, desire not to lose face and a concern for etiquette. These very constraints ensure a life without too many surprises, although perhaps a rather monotonous one.

The traditional spirit remains all powerful

The Japanese is not, as is sometimes said, a man of both East and West. The Western elements have been adopted only in order to maintain a front rank position in the world.

It is simply impossible to build vast modern buildings out of wood and using traditional techniques... Yet the Japanese draw their motivation not from the West but from their own land, their traditions of harmony, their history and the ancient religions to which they remain closely attached. They have learnt from Buddhism and from history that change is inevitable and that the spirit alone remains allpowerful.

The Japanese do not think of East and West as opposite extremes but rather as two possible ways of solving a problem. They find it easy to move from past to present, from the Eastern to the Western world. They readily adapt themselves and take advantage of new inventions. They have an astonishing ability to avoid disagreable things, they have adopted modern Western techniques—modifying them where necessary—but yet retain an essentially Japanese personality and approach. ☐

The great treasure of the Byodo in Temple at Kyoto, this Amida Buddha, a gilded wooden figure dating from the 11th century, smiles benevolently down upon the pilgrims, amid a halo of flames and musical divinities.

religion

■ Figures published in 1973 revealed that there were 81,262,600 Buddhists in Japan, divided between 200 sects, 83,074,700 Shintoists, in 200 orders, as well as 727,400 Protestants and 346,800 Catholics. This gave a total of 165,411,500 religious adherents —not bad for a country with a population of 112 million! The Japanese finds no difficulty at all in being both a Buddhist and a Shintoist; it is another aspect of the duality of the national character.

Are the Japanese religious? To the outside observer their lives scarcely seem so. They spend their childhood in a struggle to pass examinations, as adults they strive to get a job and keep it; there seems scarcely a moment for religious reflection amid all this unremitting work.

During the week social pressures oblige the Japanese to spend their leisure in company. As soon as they achieve financial security they feel they have earned a right to enjoy life. The average Japanese citizen today seems a thoroughgoing materialist, keen to enjoy his hard-earned pleasures and hardly likely to be worried about the state of his soul. Yet they often claim to be religious and to care about the ancient religions of their country. Is this merely a love of tradition? The typical Japanese is a Shintoist in his lighter moments; he tends towards Buddhism later in life—in retirement perhaps, when he has more time to think about the chances of life beyond the grave. At any rate in recent years new sects have arisen. The most important, called "Sokagakkai", has 15 million followers.

Nevertheless, it is Shinto, Confucianism and Buddhism that have characterised Japanese civilisation, given it its originality and its soul. All three have influenced the thoughts, feeling and actions of the Japanese down the centuries, inspired their literature and art and formed their institutions. Thus every Japanese today, even though he may not be an active worshipper, is deeply imbued with certain beliefs: love of the family, of nature, of art, of cleanliness, respect for his ancestors; many

Most Japanese festivals are religious in origin. The Aoi Matsuri, at Kyoto, commemorates offerings made to appease the gods during a famine many centuries ago.

aspects of his daily behaviour reflect his unconscious adherence to these moral and religious precepts.

Shinto, or "kami no Michi" (way of the gods), the earliest religion of Japan, is the worship of the "kami", the spirits of the dead. The "kami" still move among the living, determining natural events, bringing life and fertility as well as tragedy and catastrophe. The "kami" keep the character they had during their lifetime. There are good and bad "kami"; they are kind to those who bring them offerings and remember them, hard on those who forget them.

Thus the dead and the living are inter-dependent: the dead have need of the care and respect of the living, the living are subject to the protection or the hostility of the "kami". There are family "kami", venerated at family altars, village "kami", and "kami" for the natural world—sky, earth, rivers, trees and flowers. There are no less than 800 myriad "kami". In the 8th-century "Kojiki", the Bible of Shinto, there is an account of the whole Shinto mythology and of the birth of the gods and of the Isles of Japan.

The Shinto pantheon

• When the earth emerged out of chaos it was like a "patch of floating oil, a drifting jellyfish" (Kojiki). From it arose the three gods of sky, earth and water who created Izanagi (the "male inviter") and Izanami (the "female inviter"). Charged by the gods to create the world, the couple stood on a celestial bridge and beat the sea with a spear. The droplets of water that fell from their spear gave birth to the various Isles of Japan. The couple descended to earth where Izanami "invited" her brother to make love. Their union gave birth to abnormal offspring, among them an island of foam... This was the woman's fault since she had spoken first and seduced her brother. They had to start again. Izanagi cried out: "Oh, what a lovely girl! This time Izanami gave birth to handsome children: the Isle of Awaji... and the nature gods. Izanami was so dreadfully burned when the fire god was born that she died. Like Orpheus, Izanagi went down into the underworld to look for his lover. Disobeying divine instructions, he found only the hideous corpse of his beloved. He returned to earth and purified himself in a river. From the waters in which he had bathed arose Susanowo, the god of storms, Tsuki no kami, the moon-god, and Amaterasu the sun-goddess. As a joke Susanowo dropped a flayed foal down into the room where Amaterasu was sitting at her weaving. She was furious and proceeded to shut herself up in a cave, plunging the world into darkness.

The eight hundred myriad gods met in council on "the dry bed of the river of night" (the Milky Way). They decided on an appeal to feminine coquetry, curiosity and jealousy; they displayed a beautiful necklace and a huge mirror in front of the cave and bade Uzume, the goddess of laughter, to dance there. She did so in such a comic way that the gods shook with laughter. When Amaterasu asked what all this hilarity was about, they said they were rejoicing because a most beautiful goddess had just appeared. Amaterasu, filled with curiosity, emerged from her cave, saw herself in the mirror and came closer to investigate; the gods barred the entrance to the cave with a rope made of straw. Light had returned to the earth. Amaterasu decided to hand over the islands to one of her descendants and entrusted him with the three divine attributes of power: a mirror, a necklace and a sword. According to legend, Jimmu, the great-grandson of Amaterasu, became the first Emperor of Japan, in 660 B.C.

Shinto to-day

Official Shinto, introduced after the Meiji Restoration of 1868, proclaimed that the Emperor was himself an infallible and irreproachable god. This official doctrine was taught in schools. In 1945 the Americans obliged the Emperor to abdicate all claims to divinity. Yet he remains today the supreme authority of Shinto. Shinto has no rules or strict precepts, it is a body of tradition from which a family and group morality has been derived, as well as a desire for individual purification. Shintoists continue to practise ancestor worship, and their homes contain small altars or "kamidana" for this purpose. Shinto stresses the importance of hierarchy and discipline within the family, also the necessity of having a male heir.

Although Japanese family life has changed considerably—vast families no longer live together under one roof all subject to the authority of the family head—the tradition of caring for parents remains, as do the family responsibilities of the eldest son on the death of his father. Where there is no male heir, a son-in-law will often abandon his own family.

To honour the "kami" Shinto ordains the strictest rules of cleanliness; the daily bath—taken either at home or in a public bath-house—is an important leisure hour for the whole population. It is extremely dishonourable for a Japanese to neglect his personal cleanliness. This habit affects the country itself which often seems almost unnaturally tidy.

It is in the countryside, among the gods of the rivers and ricefields, that Shinto retains its greatest vitality. In the little village temples they worship the gods of rainfall and harvest; there is scarcely a hamlet where the annual Shinto festival, the "matsuri" is not celebrated. The tilling of the soil and the rice-planting are accompanied by rituals and exorcisms. There is said to be a miniature rice-field in the Imperial Palace, over which the Emperor himself recites prayers to the gods for "abundant harvests of grain, lush vegetation on the marshy plains... on the blue plains of the sea, fine-finned fish, and rich seaweeds in the depths and along the coasts."

Shinto shrines are simple buildings, rather similar to the temples and houses of Celebes and Sumatra. Their sloping roofs are thatched or covered with pinewood shingles. They are entered through a high stone or wooden archway or "torii". The origin of this structure is obscure; according to one theory they were perches for long-tailed sacred cockerels, hence their name (tori —cockerel, i—to stand). To pass under a "torii" constitutes an act of purification. At the entrance to the courtyard there may be statues of the guardians of the temple, stone dogs or foxes (see under "Kyoto", the Inari temple).

In a sacred area of the temple, to which admission is forbidden, there is a mirror of polished metal, venerated as symbol of purity. The priests wear hoods, white or green robes and wooden shoes. There are some 100,000 shrines in Japan. Shinto is a cheerful religion. Worship consists of making offerings to the gods and reciting prayers or sacred formulas. Dances or "kagura" are performed by girls to commemorate the dance of Uzume. Visitors clap their hands and pull a rope to ring a bell and summon the gods.

Confucianism

Confucianism was the first foreign religion to enter Japan, but until the 18th century its influence was limited to certain educated circles. Confucianism is more a philosophy than a religion. Why should men worry about heaven, for they can know nothing about it. Earthly things are just as important; they should pay attention to them. The dead should be venerated without asking questions about survival. This down to earth approach suited the Japanese temperament, while the conservative family ethic proved congenial to a people whose spirit had been so much influenced by Shinto devotion to the family and to the past. Confucianism commended the veneration of one's ancestors. It stressed the importance of courtesy, filial piety, as well as respect for those in authority, one's masters and superiors. The superior has the rights of a father; this idea of family hierarchy is still apparent today in the relationship between employer and employee in big Japanese

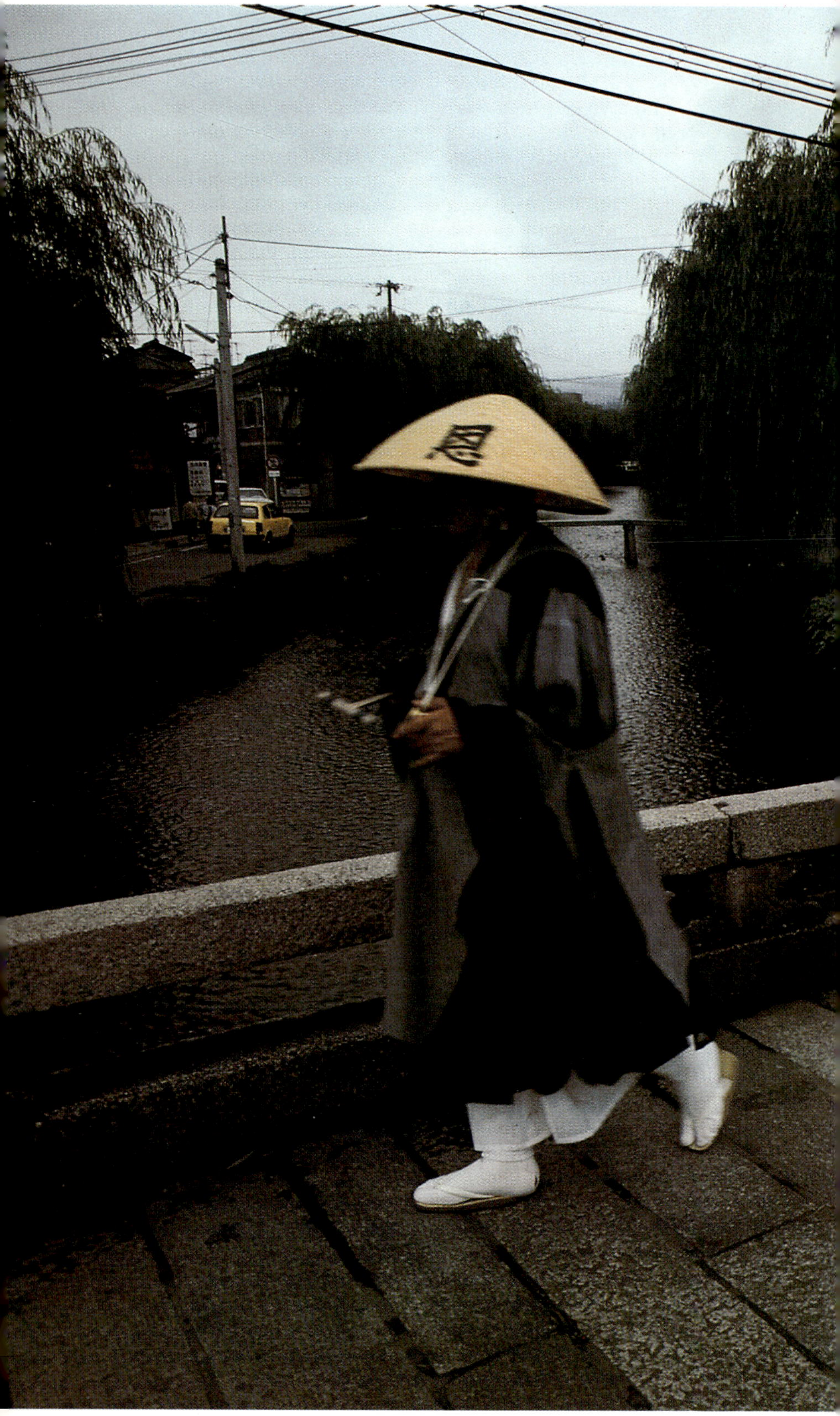

The Japanese are often Shintoists for most of their lives, turning to Buddhism later on. At all events these beggar monks are unlikely to go hungry.

firms. Confucianism advocated knightly loyalty, unto death if necessary. If a master had suffered an insult then he had to be avenged; in case of failure it was fitting to commit suicide by ritual disembowelment, "seppuku" or "harakiri" (see special entry: *the 47 ronins*). Feudal military ethics, "Bushido" or "the way of the warrior", were imbued with Confucianism. During the 17th century, the shogun Iyeyasu had the Confucian classics published and distributed to all classes of society.

Buddhism

When Buddhism arrived in Japan from Korea in the 6th century it ran strongly counter to the Shinto national consciousness. Buddhism is a metaphysical system. This present life is one of suffering (sickness, old age, the instability of things, the inconstancy of feelings). The desire to exist lies at the root of this suffering. When he dies, an individual becomes reincarnated and this reincarnation will be an improvement, or otherwise, according to whether his life has been a good or a bad one. "Nirvana" is attained by overcoming one's own desire for existence. For ordinary people "Nirvana" is the Paradise of Amida. The early Buddhists were persecuted by the Shinto priests. Shinto acknowledges a vast number of gods. For the Buddhists the gods play no role in this world; Buddha himself was not a god but a bringer of enlightenment. Shinto holds that the spirits of the dead survive without punishment or reward, a doctrine quite contrary to the Buddhist belief in the transmigration of souls. In order to gain acceptance, Buddhism had to compromise. The great gods of Shinto were held to be incarnations of the Buddha. The spirits of the dead were now believed to survive alongside the living, for a certain time, before being reincarnated. Buddhism quickly became very influential. During the Nara period it was associated with the Chinese culture and learning of the monks. Before the building of a temple the Emperor would "warn" the goddess Amaterasu so that she would not take offence. Soon Buddhism came to be regarded as wisdom to be striven for, and life as the hard school in which it was learned. The Japanese adapted Buddhist funeral ritual into a ceremony for appeasing the spirits of the dead. In the 9th century the monk Saicho founded the "Tendai" sect which maintained that every individual possesses the nature of the Buddha. Its monastary, on Mount Hiei overlooking Kyoto, became over the years a centre of "soldier monks" who had to be crushed during the 16th century. At that time also "Saint" Kobo Daishi founded a temple on Mount Koya as the home of the "Shingon" sect. They asserted that through art alone could man come to know perfection. Over the centuries the "Shingon" sect were to contribute greatly to the Japanese artistic heritage.

During the Kamakura period (late 12th to early 14th century) culture ceased to be the privilege of the élite and Buddhism spread to all classes of society. The monk Shinran founded the "Jodo-shishu" sect (not to be confused with the "Jodo", founded by Honen, the initiator of Shinran); they held out the promise of paradise to any ordinary man who prayed to Amida. The monks traversed the country bringing comfort to the poor. This sect was so successful in bringing its message of peace to the people that, when the Honganji temple in Kyoto was burned down, the peasant women cut off their hair and gave it to be made into rope to help rebuild the roof.

"Zen" Buddhism proclaims the virtue of silence. "Seek within yourself the virtue that you cannot see." Enlightenment cannot be attained by prayer, study and discussion, but by intuition. The Zen master's questions to his pupil do not admit of logical answers. "Zen" became the religion of the military caste who had a taste for the absolute but were not inclined to meditation. "Zen" was introduced into Japan by an Indian monk, Bodhidharma, who meditated for so long standing up that he wore his legs away. He is depicted as a big sphere with two painted eyes and a nose. He is known as "Daruma". In the 13th century a monk called Nichiren revealed himself as the new Buddha "for whom the world has been waiting for so long." He was considered a trouble-maker by the other sects and was exiled time after time. His followers were crushed during the 16th century but a small group survived. During the last war they were imprisoned for refusing to acknowledge the divinity of the Emperor, since they possessed the "true" Buddha. The "Sokagakkai" sect (society for the study of the creation of values) of today claims to follow the doctrines of Nichiren and is becoming very influential. It has some 15 million fanatical supporters. The Sokagakkai recruits its followers mainly among women and the downtrodden. A sort of freemasonry exists among them and they help any member who may be in need. As well as mutual aid they are committed to abjure Shinto observance. In the cities there is a meeting hall and a school in every district. The sect publishes a newspaper which has a circulation of three million; it supports the nationalist political party known as "Komeito".

The Buddhist temples are built of wood, painted, lacquered and often carved. They usually have tiled roofs with slightly upturned eaves. They consist of a shrine for the veneration of the Buddha, a "pagoda" several stories high, a covered gallery round the main buildings, a library, a belltower, and dining halls and pavilions originally built for aristocratic retreats. The temples house fine statues of Buddha Amida, and of Kannon the goddess of mercy. During the services the priests read and sing.

In Japanese Buddhism there is "baptism", "ordination", and "confession" for the monks; funerals are accompanied by elaborate ritual. The living honour their ancestors by leaving offerings on the family altar.

The great festival of "O Bon", which takes place in July, celebrates the return to earth of the souls of the departed who come back to visit the dwellings of their descendants. In the villages white lanterns are lit at the doors of the houses, to guide these returning souls. The houses themselves, scrupulously clean, are decorated with flowers and food. Monks go from door to door chanting prayers. Fishermen wait on the shores to guide the souls on their journeys.

Catholics and protestants

It is only during the past hundred years that Catholics and Protestants have been able to conduct missionary activity among the Japanese (see chapter: *Japan through the centuries).* They have not yet had great success. It is not easy to convert the Japanese. It is not so much on strictly religious grounds that they are unwilling (they could easily add Christianity to the two faiths they hold already) but rather because of the revision of values involved—the role of the individual in society, the position of wives, and the attitude towards one's neighbour.

The Japanese, who has struggled since childhood to make his way in life, wonders what Christianity can do for him, what use it will be. Moreover, the judgment and opinions of his colleagues complicate life for a convert. However, Japanese women are sometimes attracted by Catholicism; they feel it accords them a value they do not always have in their traditional faiths.

Yet the Japanese, as always, take advantage of the positive aspects of a new situation. Both Catholics and Protestants offer excellent education to children and students, and they have very good teachers. The primary schools run by the Salesians are well attended, as is the Sophia University. There are even a few Jesuits and Dominicans teaching philosophy and literature in the Japanese state universities. These missionary-lecturers speak fluent Japanese, having themselves studied alongside Japanese students in universities. They know the mentality of the people and how to capture the interest of their audience—an essential first step. □

the life of mr. yamagata

■ According to ancient custom, theoretically abolished at the end of the war but persisting still in the countryside and in small towns, Baby Yamagata is already one year old at birth. As all Japanese add a year to their age on 1st January, it can happen that a child born on 30th December is 2 years old just two days later! During the ceremony of "Nazuke", which takes place a week after his birth, Baby Yamagata receives his name. The choice of this name, an auspicious augury for his future, is often a difficult process.

The child is king...

When the child is thirty days old his parents take him to a Shinto shrine (no matter whether the parents are Buddhist or Shintoist, this ceremony, a joyful occasion, is a Shinto one) for him to be welcomed by the community. This is known as "Miya mairi." His mother, dressed in a gorgeous kimono, offers the baby, wrapped in a splendid robe, to the priest who waves over him a wand decorated with paper ribbons, a sign of purification. Nowadays the ceremony usually takes place on a Sunday. If it is raining the mother will arrive dressed in Western clothes and change at the shrine into her kimono. Afterwards presents are exchanged between the parents and their friends and they all sit down to a good meal.

The baby will be spoiled by everyone at first. His links with his mother will be very close; she will carry him on her back for the first three months and he will bath and sleep with her throughout his early childhood. Up to the age of six he will be able to do just as he likes. Such little Japanese children, treated like kings by their families, are often unbearable as they shout and chase each other round shops or hotel foyers—always under their mothers' admiring gaze. In trains you often see them sitting while adults stand. Twice a year great festivities are held in their honour. In March the little

*The Japanese are keen photographers;
festivals give them plenty of scope!*

girls are queens of the Dolls' Festival. For the Boys' Festival, at the beginning of May, great carp (a fish noted for its longevity) made of coloured cloth are hung from poles in front of the houses—as many carp as there are boys in the family. Private rooms are reserved in hotel restaurants for special celebrations in their honour.

When he is about four years old, young Yamagata will enter a kindergarten. The choice of this first school is extremely important, for it is said to have a bearing even on university entrance. Good kindergartens are privately run and the fees are about 8,500 yen a month. To this must be added the entrance fee of about 30,000 yen. From now on, parents who have any ambition for their children have to save all they can from the father's wages or monthly salary. Over the years they will spend a fortune on their son's education, until his entry into a university, sending him to the best schools and hiring private tutors to coach him for examinations between classes and for public competitions.

School and university

In Japan there are nine years of compulsory schooling. At the age of six, Yamagata starts the first of six years at a mixed "Shogakko" (the equivalent of our primary school). During this time, every weekday (Saturday included) from 8.40 to 4.30 (the mid-day meal is taken at school) he will learn to read and write the 1,800 Sino-Japanese characters needed to read a newspaper, to use brushes, to calculate, using the traditional abacus or "soroban."

At the end of these six years he will also be able to read Japanese in Roman lettering, as a basis for learning other languages, especially English. Sport has an important place in his education. Yamagata has three lots of school holidays: a month during the hottest part of the summer, and a fortnight in the winter and before the beginning of the school year in early April.

At the age of twelve, Yamagata begins the first of his three years at the "Chugakko" (secondary school). His pampered early childhood is a distant memory now. He has to work unremittingly in order to pass terminal examinations in all subjects. From this age on, 60 per cent of Japanese children have a private tutor to help them with their English, mathematics and so forth. They live under constant tension and many suffer from precocious stomach troubles. Japanese doctors are used to coping with this; there are children's wards entirely filled young patients suffering from these illnesses in the large Japanese cities. While in hospital they can often follow special courses so that they are not too far behind when they finally go back to school! From the beginning of his schooldays, Yamagata is taught how to live in a group. His teachers and fellow pupils initiate him into the art of being accepted in his class, of joining in with his colleagues and working in and for the group, so that it will be the best. Several times a year the class goes out on educational excursions—you see them standing lined up, in uniform, attentively watching, learning and noting the explanations given by the teacher.

A hard entrance to a college

At 15 compulsory schooling is over. Yamagata decides to continue his studies. He will try to pass the difficult entrance examination to a "Kotogakko" (State college). If he fails he will try the easier one to a private college. But in either case his education will be far from free. Pushed by his teachers and tutors, with the moral and financial support of his parents, Yamagata will work himself almost to the point of a nervous breakdown, to prepare for university entrance.

Until 1868, education in Japan was the preserve of the nobility, the military, and the wealthy merchant classes who attended the Buddhist private schools. The court abounded in educated women. At that time only about half the population received any education at all. Under the Meiji reforms an educational code, based on the French system, was introduced in 1871. Four years' schooling became compulsory for all, but it had to be paid for. Free primary schools were not introduced until 1900.

In 1947, under American influence, compulsory schooling was extended to nine years and all Japanese wishing to enter a university had to undergo three year's preparation.

The Japanese University

In 1855 the first Japanese university opened its doors, in Edo (Tokyo); it is known today as the "Todai". Other state and private universities followed, including "Waseda" in 1882, and "Keio" in 1890. Before 1947 there were 47 universities in Japan, serving a population of 70 million. Today the population has reached 100 million and there are 900 public and private universities. Four out of ten pupils pass the examination to enter a college at the end of their compulsory schooling; one in four enters a university. There are two main categories of university; those offering two-year courses (480 in 1970) generally attended by women, and those where courses last four years (382 in 1970), here 80 per cent of the students are men.

In 1971 there were 1,700,000 university students, 21 per cent of the population aged 18-22 years; in 1976 there were 2,157,000. The Ministry of Education plans to raise the percentage to 50 per cent by 1980. The State finances the universities to the extent of 95 per cent. The remainder is covered by students' fees (roughly 35,000 yen a year). Private universities receive very limited state support and fees are high (roughly 300,000 yen a year); medicine is the most expensive course. To these figures must be added entrance fees ranging from 60,000 to 1 million yen, according to the faculty.

Yamagata as a student

Throughout his three years in college, as a "ronin" or "soldier adventurer", Yamagata takes examinations right up till D-day. In order to increase his slender chances of acceptance, he takes entrance exams at various universities in succession. During a single week he will present himself on Monday at the Todai, the state university in Tokyo; on Tuesday, at the Sophia, a catholic university, also in Tokyo; on Thursday, in a specially-chartered plane, he will set of with other candidates for Sapporo, to take the examination there, on Friday, somewhere else...

Yamagata is a good candidate, but he is by no means certain to obtain a place. Out of 650,000 candidates for entrance to the various Japanese universities in 1972, only two thirds were accepted. Every year the Todai University takes 2,000 of the 16-20,000 who present themselves, while the Sophia Catholic University admits only 1,800 out of the 25,000 who take the examination.

Yamagata passes his entrance examination. Not all his friends are so lucky. Spring, season of examinations, is also the season of student nervous breakdowns and suicides. Why do some 300 young people commit suicide every year? The family of an unlucky candidate has deprived itself for fifteen years. It had pinned all its hopes on this son; now he has disappointed his family group, failed in what they expected of him, he is a loser in the inevitable race which anyone who wishes to gain a higher social position or a job in a big firm has to enter. He has failed to master the situation. His family group did not except him to be the most intelligent or the ablest, but to succeed, to be the strongest. He has lost his place in society, so he leaves, he kills himself.

Having passed the examination Yamagata can now relax for four years. For those who have painfully made their way to the entrance to a university, the exit is not so difficult. They are almost certain to obtain their final degree. Their university years are a period of relative rest and calm, to prepare them for the life of struggle that lies ahead for most Japanese. After finishing a first degree it is possible to spend a further five years studying for a doctorate.

Mr. Yamagata's professional and private life

Despite their fees, the private universities are rated highly by the Japanese for their excellent teaching, largely in the hands of Jesuits and Dominicans.

However, the top civil servants are usually the products of the large state universities.

For fifteen years now, Mr. Yamagata has been working for Mitsubishi. On leaving the university he was successful in the entrance examination to a prized position in this big firm. His starting salary was 80,000 yen (£ 165) a month, plus 680,000 yen a year in bonuses. These bonuses, paid usually twice a year, can amount to as much as seven months' salary. Smaller firms, employing only about fifty workers, offer correspondingly smaller bonuses—often only 160,000 a year. It is the dream of every Japanese to get a job with a big firm, because of such advantages.

When Yamagata began working at Mitsubishi he could no longer live with his parents in their small flat, so the firm found a place for him in one of their hostels. His studies gave him accelerated promotion through the lower grades of the firm's administration. When he wanted to get married his manager gave him a loan so that he could rent a 4 ½ "tatami" (rice straw mat, used as square measure in Japanese houses), one-roomed flat; he also found a nice wife for him, exactly what he wanted. She had brothers who would guarantee that her family would not one day become a liability to him. For Yamagata, as an "elder brother", knows that on the death of his father he will have to take moral if not financial responsibility for his mother and his younger brothers; he will succeed his father as "head of the family." Admittedly this is not such an onerous business today as it used to be when the whole family lived together under one roof, yet certain responsibilities remain; it's not by accident that the Japanese ideogram for "elder brother" consists of a square and two legs, symbolising an active man! Yamagata did not risk being "adopted" by his fiancée's family, as is sometimes still done in families where there is no male heir.

He is a typical "medium grade wage earner", a category which covers 55 per cent of the Japanese population. He lives in the outer suburbs of Tokyo, an hour's journey by underground from his office. Yamagata gets up at 6 o'clock every morning and dresses with care: plain neutral-coloured suit, impeccably laundered shirt, tie. His breakfast, prepared and served by his wife, consists of a raw egg which he breaks into a clear soya-bean soup, rice and dried sea-weed soaked in the soup, and a bowl of green tea... He leaves the flat and a few minutes later is caught up in the struggle to board an already crowded train—there are "pushers", people (often students) specially employed to pack the passengers into the carriages. Tight-pressed but stoical, Yamagata waits for the hour's journey to pass.

No job change
No holidays

He wears his suit jacket all the year round inside his air-conditioned office. Only during the hottest days of summer, when he goes down to the snack bar with his colleagues, to lunch, does he leave it behind in the office. The hosts of men who invade the streets at such times, all in their white shirts and ties, look like so many schoolboys in their Sunday clothes...

The years go by Yamagata, now aged fifty, has a new air-conditioned flat—two rooms plus kitchen and bathroom. He has bought a colour television, a washing machine and a vacuum cleaner. As a divisional head

*Sunday is the day for visiting the temples.
Parents and children purify themselves
amid the smoke of a thousand incense sticks
and then, their duty done,
go off to enjoy themselves
at the little shops and stalls nearby.*

he now earns 150,000 yen (about £320) a month, plus bonuses, and pays taxes amounting to a quarter of his salary. If he is ill there is the Mitsubishi hospital. His sons attended the excellent kindergarten run by the firm. He is pleased with his job and intends to stay in it until he retires. In Japan they do not think well of people who change their jobs, so Yamagata does his best to remain on good terms with his colleagues and with his boss, in order help his group contribute to the smooth running of the larger family, the firm.

For fear of losing his job, Yamagata has never taken a long holiday, but four times a year he spends a week-end fishing with cormorants at Gifu or Matsushima.

When he retires he plans to travel a bit, perhaps to Hong Kong, if he has not spent all his money on his sons' education, or on travel for them. He is ambitious for his children and would give up anything for their sake.

At midday, Yamagata leaves his office with a few colleagues to breath a little Tokyo air and lunch off a "soba" and a few "sushi" (see: *Japanese food*) in a nearby restaurant. Like all employees of big firms, he has an "open account" (tax free) for such expenses as restaurants, bars, night clubs perhaps, a subscription to a golf club and the expenses of business travel abroad; he pays for such things by cheque in the name of his firm. This expense account sometimes totals as much as the annual earnings (apart from bonuses) of an employee.

Patchinko: a game for a whole population

Often, after his day's work at the office, Yamagata will not get home before midnight. He may go to the public baths in the neighbourhood and then perhaps attend a business dinner with "geishas" (see: Kyoto). More often he will go out with his colleagues. It is important for him to be on good terms with his office group, to return home early would be a sign of deplorable individualism! His colleagues would probably think he was in his boss's bad books. So Yamagata goes out with the rest of them.

They go to bars, and play "mahjong" (a sort of Chinese dominoes) and "patchinko".

Patchinko was invented in Osaka during the dreary post-war days. A Korean had the idea of using war surplus ball-bearings in a game based on the pin-ball machines at that time very popular in the United States. It was a tremendous success. There are some 11,000 patchinko halls in Japan, 4,000 of them in Tokyo; 70 per cent of the population play the game regularly. Almost every evening Yamagata settles down in front of what looks like a "flipper" machine. He never wins any money, his three hour game which costs about 2,000 yen brings him a few sweets and some cigarettes, at most a small transistor radio. Some Japanese spend as much 37,000 yen a year on the game.

Before or after their patchinko Yamagata and his colleagues will have dinner together. Ten o'clock will find him in a bar, chatting to one of the 250,000 "hostesses" employed in Japan, like mini-geishas, to talk to customers, put them at their ease with chat about the latest films or make them laugh at the latest joke and so encourage them to spend... When the bars close at 11 o'clock, Yamagata will leave, staggering slightly perhaps, supported by his friends in a similar state. They will all make for the underground.

The Japanese drink relatively heavily and drink tends to affect them. According to statistics, the average consumption per head is 13 litres of sake, 83 litres of beer and 5 litres of whisky a year. Being drunk is not considered dishonourable, but thought quite normal. The Japanese usually drink with their work group, to reinforce its cohesion; they rarely drink alone.

Towards midnight, Mrs. Yamagata, who has been sitting watching television, will be joined by her husband, off the last train. He usually spends only seven hours out of twenty-four with his wife—the nights in fact—he spends the rest of the time with his colleagues and in travelling. During the week he never sees his children.

When he is fifty-five Yamagata will retire. On retirement he will receive a sum equivalent to three years' salary. In addition to these several million yen his firm will pay him a small annual pension, for eight years. He also has the opportunity, if he wishes, to take a temporary post with the firm, for six years, at a third of his previous salary.

During the remaining fifteen years or so that remain (the average life expectancy in Japan is 71 years for men, 76 for women), he can relax, enjoy his modest savings, watch the plants growing on his balcony and reflect on his busy past life.

What sort of a life was it? As a schoolboy and a young man Yamagata had to study all the time to pass examinations. Later on he was lucky enough to get his job in a big firm where he spent all his working life, making his modest contribution to the success of his group and the enterprise as a whole.

On Sundays he used to go out in his car (three quarters of the people own a car—if they can get a garage or a parking place) with his wife and children, or play golf, or go to the seaside in the summer. His somewhat humdrum life was brightened up by the various festivals, good meals, an occasional theatre and a week-end break. Does he have regrets? Looking back, none whatsoever. Fate ordained this life for him and he fervently hopes that, in this overcrowded country where jobs are such a problem, his own sons will be lucky enough to enter a big firm like he did, and enjoy the life of an "average" Japanese. □

JAPAN: A POLLUTED COUNTRY?

■ *After the Second World War, Japan bent her energies towards becoming a major industrial power. There were, of course, the inevitable byproducts of industrial fumes an industrial wastes, spewed out into air and water. As it became evident that these concomitants of industry were causing organic diseases, their effects were carefully studied and two laws were promulgated, on atmosphere pollution and clean water. The National Environment Agency (1971) concerns itself exclusively with limiting chemical pollution from factories and from motor vehicles. The latter have to keep to strict limits as to their production of carbon dioxide and hydrocarbons. The police carry out spot checks on vehicles in the streets. Motor plant assembly lines now include instruments to check a vehicle's fume production. The factories themselves are obliged to reduce their own output of fumes—especially of sulphur fumes which used to be a major cause of chest complaints in Japan. Thanks to such actions, atmospheric pollution is considerably less than it used to be, in Tokyo and other main industrial centres. At the moment there is a major campaign to limit the emission of hydrocarbons and other causes of smog. The law on clean water limits the amount of chemicals, particularly mercury, that can be disposed of into rivers and seas. Already by 1974 the annual spending on pollution control had reached a thousand billion yen; the campaign is being maintained.*

*The important stages of life
are marked by a visit to a Shinto shrine.
Here a month-old baby, sumptuously robed,
is presented for the ceremony of "Miya mari".*

the japanese woman

■ Dressed in her kimono, trotting along in her sandals, "getta", or pivoting on her high heels, wearing the light-coloured materials of which she is so fond, Yuki, the future Mrs. Yamagata, seems the very incarnation of femininity and charm. Like a butterfly emerging from its chrysalis she has left her dreary adolescence behind her. Her schooldays over, all her efforts are now directed to making herself attractive. For Yuki and the other young women of her age—university students, office-workers, secretaries and shop assistants —are all aiming at the same thing, getting—and keeping—a husband. In present-day Japan only two per cent of women aged thirty and over are unmarried. How will Yuki spend her time until her marriage? First of all she will finish her secondary schooling, like some 83 per cent of Japanese girls. Then she will enter a women's university, where she will learn how to run a household as well as complete her general education; or she may take a job. By the age of twenty-four, at the latest, she will expect to be married! In Japanese factories women earn half a man's wages. If they work at home they receive only a quarter of what a man would earn for the same work. One wonders whether any Japanese today recall that, until the 7th century, the country was matriarchal? It was not until 1924 that women had the right to hold public meetings; on 3rd May 1947, thanks to the Americans, they acquired the vote. If she takes an office job, Yuki will have to make herself look pretty and dress most carefully. It will be up to her and her colleagues to provide a pleasant atmosphere for the men of the firm to work in.

Until quite recently, women employees had to arrive at the office a quarter of an hour before the men, in order to arrange the flowers.

Will Yuki be able to marry any man she pleases? If she lives in Tokyo or another big city, it will be easier than in a provincial town or in the country, where marriages are often still decided without consulting the parties; even in Tokyo, meetings

When the geishas are trotting off to their first evening appointments the patchinko parlours are opening to admit their eager customers. They will spend several hours playing happily here, forgetting their cares for a while and perhaps winning a few sweets or some cigarettes—never money.

are often "arranged". It is still very rare for a couple to marry in defiance of their parents' wishes. In every town there is an office in the Town Hall where information is available about the families of those about to be married.

A somptuous feast: her marriage

Thanks to an intermediary Yuki has met Mr. Yamagata. She is twenty two. The great day of her wedding has arrived. Her family, like most families in Japan, will spend lavishly for the occasion (500,000 yen, almost £ 1,000, on average) which will be celebrated in a more or less luxurious hotel, according to her parents' means.

In Japanese hotels there are large rooms reserved for weddings, with often an adjacent shop where kimonos can be hired at various prices, also accessories for the bride and the couple's relatives. The shop will also stock suitable presents for the guests. Before her wedding, Yuki will visit it and choose, from a selection of wax models in the window, boxes of seaweed, biscuits or preserved exotic fruit, which she will give to her guests. Since the cost of a kimono is often exorbitant (sometimes from 3 to 10 million yen), Japanese brides usually hire them (they consist of three garments worn one on top of the other) as well as the hat they will wear on their wedding day. It costs 200,000 yen to hire a fine kimono, with another 100,000 for the "obi", the broad belt worn with it, for the three hour ceremony. The actual dressing of the bride, for which a professional dresser is hired, usually costs 25,000 yen. Kimonos for the bride's parents will add another 10-60,000 yen to the bill.

A Buddhist ceremony takes place in a temple, a Shinto one may be held either in a shrine or the priest may come to the hotel and hold it there. After her marriage, Yuki will give presents to her guests, who will offer theirs to her; they often consist of an envelope containing money, thus lightening the cost of the proceedings for her parents.

Having acquired a husband Yuki now has to look after him. When her first child is born, and if finances permit, Yuki, like most Japanese wives, will cease going out to work. Japanese husbands take a certain pride in being the sole breadwinner of the family. Mrs. Yamagata looks after the housework, deals with the schooling and education of her children and looks after the family finances—her husband hands over his entire salary to her, apart from the famous monthly bonus.

Neither her husband nor children come home at midday; a Japanese home is easily looked after and ordinary Japanese food is simple to prepare. Yuki has plenty of free time; she goes out, sees her friends and usually watches television for three or four hours every day. Hers might seem an ideal existence. Japanese wives enjoy a freedom of action and decision that many Western ones might envy. But what does she mean to her husband? He certainly respects her, she is the life and soul of his household and the mother of his children. Like all Japanese, Mr. Yamagata makes a clear division between his business and his private life, he looks upon his home as a peaceful retreat from the world outside. His wife, with her gentleness and charm, knows the secret of how to provide this essential calm. As an engaged couple, both of them students, they had many interests in common, but they knew that when they married each would live their own life. As man and wife they are together in the evenings, though often not until late; they don't talk to each other very much, but this doesn't seem to matter, they learn to understand each other without words. A Japanese rarely tells his wife about his work or about what he thinks. Sometimes he brings his colleagues home, but most business dinners take place in restaurants where "geishas" act as hostesses.

This is a normal accepted way of things. There is no feeling of superiority or inferiority between a Japanese maried couple. Each occupies a position ordained by centuries of experience. Japanese women are brought up to play a self-effacing role. Questions such as whether they find such a role pleasant, satisfying or adequate, remain unanswered. "Shi kata ga nai"—"there is nothing to be done about it"—is a common Japanese phrase, though one wonders whether all Japanese women believe it. At any rate those involved in advanced studies, or with interesting jobs, are tending to marry less or later, thus deferring what they see as a less fulfilled life.

But time is passing by...

Mrs. Yamagata knows that her beauty will fade with the years, so she uses all the means at her disposal to retain her youthful charm for her husband. She instinctively chooses soft dress materials to achieve this effect, wearing a silk kimono in the winter and a fresh "yukata" (cotton kimono) in the summer. Thus she enhances her gentle, captivating fragility. Yet, as her children grow up, and with her husband often away, she gradually slips into old age—travelling perhaps or helping in the nearby temples.

In the countryside a wife as well as running the house will often help her husband in his farm work. You see her in the paddy-fields, ploughing and planting out the rice... A hard life indeed, but one which perhaps brings the couple together more than life in the towns.

On the whole the Japanese woman of today lives, like Mrs. Yamagata, somewhat in her husband's shadow, but firmly in control of the running of her home. □

THE DOLL FESTIVAL AND THE CARP FESTIVAL

■ *During the 18th century it was the custom for the children of the nobility, once a year, to offer dolls to the daughters of the rulers of the land; these dolls were supposed to attract to themselves any diseases which might otherwise have afflicted their owners. After the Meiji Restoration this custom extended to all classes. The festival of "Hina-Matsuri" is celebrated in March. Shopwindows contain display stands draped in red cloth; the Emperor and Empress are enthroned on the topmost shelf; on the shelves below are ladies of the Court, musicians, miniature furniture, screens and bibelots. In family homes dolls—some of them real collectors' items—sit in state in the "tokonoma", alongside an "ikebana" of peach blossom. On the day of the festival itself, little girls receive all sorts of presents and invite their friends round to eat the special cakes that their mothers have made for the occasion.*
The Carp Festival, at the beginning of May, is essentially for boys. Cloth or paper carp wave gaily from masts set up in people's gardens or from their roofs—as many carp as there are boys in the family, with two bigger ones representing the father and mother. Carp stand for longevity and symbolise the strength and courage that the boys will later demonstrate in their lives. Rooms are hired in hotels for special banquets in the boys' honour. Special cakes are baked in the shape of boats and tanks. At home, in the "tokonoma" there are displays of models of the armour worn by Yoshitsune, the warrior brother of Shogun Yoritomo.

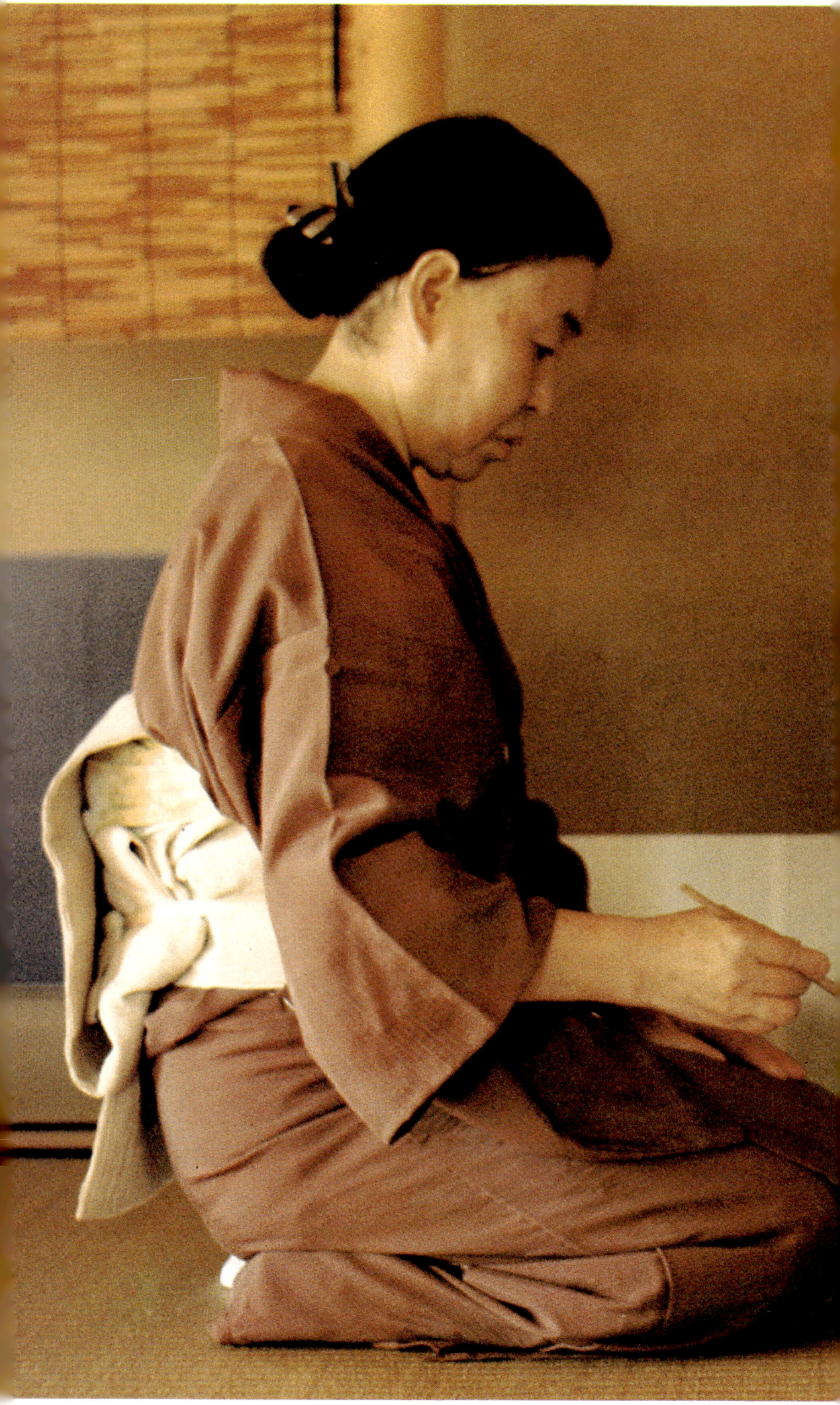

*The Tea Ceremony is a search for beauty
in colour, form and gesture,
a school of self-control.
It brings repose to the body
and clarity and detachment to the mind.*

the japanese home

■ Japan is a country of some 370,000 square kilometres, with a population of 110 million. Three quarters of her people live on the 80,000 square kilometres of plains, 58 per cent of them in the big cities such as Tokyo, Nagoya, Osaka and Kitakyushu. In the most crowded areas of Tokyo there are 14,800 people to the square kilometre.

A foreigner arriving here, hoping to find in the cities the charming houses of Old Japan, will inevitably be disappointed if he sticks to the usual route: Tokyo, Osaka, Kyoto, Kamakura... He will take away a memory of vast cities, cluttered with twelve-storey blocks and municipal housing. This would be a false impression of Japan, since it is an incomplete one, based on that quarter of the country that is most heavily industrialised. By contrast, the other three quarters present a much more traditional appearance.

If the visitor makes his way inland in Honshu, explores the little valleys that run down from the mountains to the Sea of Japan in the north, or the islands of Shikoku and Kyushu, he will discover that, as soon as he leaves the industrialised strip along the Inland Sea, the traditional picture of Japan awaits him.

Wooden houses and little gardens

The old wooden houses of Japan still stand amid their little gardens, varying slightly in style according to region, their wide upward-curving blue-tiled roofs reflected in waters of the ricefields. In the smaller towns, the streets of the older quarters are still lined with traditional wooden houses. (They last about twenty years; a four-roomed one costs around 35 million yen.) In the outlaying districts, if space permits, each modest house will be surrounded by a garden, perhaps a tiny one. It is astonishing to find peaceful cottage quarters just a stone's throw from Tokyo's main streets... Only if he is obliged to by his work in some industrial centre does the Japanese take to living in a tenth-floor flat. They are all nature lovers and all dream of retiring, at fifty, to a little house and garden.

The traditional Japanese house is built of wood and slightly raised. It is surrounded by a verandah onto which the rooms open; this verandah is either left open to the sun or enclosed by sliding wooden screens. The inner wall is made of sliding screens of wood and translucent paper, called "shoji". Movable paper partitions, "fusuma", separate one room from another and make it possible to vary their size at will. When all the screens are open the garden, designed to be seen from every room, seems to become part of the house itself.

Inside, the floors are covered with "tatami", thick mats made of rice-straw; it is normal to remove one's shoes before walking on them. At the back of the main room in an alcove, the "tokonoma", hangs a painted scroll or "kakemono" which is changed according to the season, the festival or the weather, and a restrained flower arrangement, an "ikebana". The most important guest is seated near the "tokonoma", "a place surrounded by beauty." When the family commemorates its ancestors, inscribed tablets are displayed in the "tokonoma."

The most striking feature of a Japanese home is its bareness. There is little furniture, clothes are kept in cupboards which also contain the Japanese beds, "futon", thick mattresses covered by a quilt, which are unrolled at night on the tatami. The fact they can be overheard anywhere in the house induces remarkable self-control in the Japanese, from a very early age.

An extremely simple decoration

One of the most important rooms in a Japanese house is the bathroom. Both Shinto and Buddhism regard cleanliness and purity as a rule of life. Japanese of all classes take frequent baths. If a Japanese returns home after work he will take a bath before dinner; otherwise he will queue up at the public baths close to his office, before going on to the restaurant or to play "patchinko". The traditional bath is like a small, semi-sunken swimming pool. Families either take their baths together or in order of seniority: father first, then mother, then the children. They first soap themselves all over, outside the bath, then rinse under a shower or from little wooden buckets, before immersing themselves in the hot water (40-42°C.) where they often remain for a long time—chatting with other members of the family in the adjacent rooms. This custom is a feature of Japanese life in all levels of society.

Such is the traditional Japanese home, still common in the countryside and in the smaller towns and the dream of city-dwellers; the citizens of Tokyo and Osaka usually live rather differently.

The dream in towns: a two-room flat

Japanese firms see marriage as a stabilising influence and encourage their male staff to marry by offering them, at the age of thirty, a loan to buy or rent a flat. Half the people in Tokyo own their own flats, consisting usually of a single room, plus a tiny kitchen and small bathroom; 45 per cent are 4 ½ "tatami" in area. Rents are very high. A municipal flat in Tokyo, a 4 ½ "tatami" room, costs 20,000 yen a month; a 6 "tatami" one, plus kitchen and bathroom, in a rather more comfortable building, costs 46,000 yen. At Kamakura, in the outer suburbs of Tokyo, an hour's journey from the city-centre, where many city workers live, a municipal flat comprising a living room, bedroom and kitchen, comes to 46,000 yen a month.

A young married couple usually start off by living with the husband's parents, a stage that the bride often hopes will be a brief one, as she frequently does not get on with her mother-in-law! The couple will then move into a little one-room flat. The single room will be furnished in traditional style. During the day the family will live and eat there; it's where the mother will work and the children will play. At night everything will be put away and the "futon", beds, will be brought out.

A few years later, when the man's salary has risen to about 150,000 yen a month, plus bonuses, an average couple will be able to move and consider renting or buying a two-room flat with kitchen and bathroom, near-luxury by comparison!

Such a flat is known (in English) as a D.L.K., i.e. a dining room, living room, kitchen; the dining room is usually furnished in Western style, with table, sofa and television; the living room remains Japanese and can take as many "futon" as there are occupants. It is covered with "tatami". If the flat has no bathroom the new owners will have one fitted. Since her husband often does not return home until late in the evening it is the wife who usually suffers most from this restricted living space.

By the time the husband is fifty the couple may well have spent so much money on their children's education that their retirement dream—a little wooden house in its own garden—will prove beyond their means... □

Painting expresses the essentials of Japanese art. Greatly influenced by their training in calligraphy, Japanese painters possess the skill of rendering their subjects in fluid, economical brush-strokes. (18th-century painting, in the Todai ji at Nara)

the arts in japan

■ In Japanese art we see a constant battle to extract the essence from the medium. It is always a matter of isolating this essence and giving it its true personality. In all the arts of Japan a love of nature, of plants and of the medium itself is present and visible. The artist succeeds in rendering nature as it really is, with all its changes of mood.

The Western vision often feminizes nature, a landscape is often more an expression of the artist's personality than a representation of the thing itself. There is no romanticism in the Japanese artist as he lovingly depicts nature in all its beauty, its changeability and even its ugliness. Japanese art is not pretentious; Japanese craftsmen are modest, they do not consider their medium as something to be "dominated" but rather as "an aspect of life, to be understood and loved", they approach their medium with instinctive sensitivity.

Geographically isolated, Japan was for a long while dependent on China from whom she received, somewhat delayed, her artistic impulses. It was from China that Japan took her understanding of space, the place of man in the natural world, as well as the very materials of her art: bronze, lacquer, bamboo, silk and paper.

"Haniwa" and the great tombs

Clay statuettes have been discovered dating from the Jomon period (3000-300 B.C.); they represent human figures, broad hipped and with wide-open eyes. During the 3rd century B.C., in the Yayoi age, these statuettes give way to bronze bells and pottery turned on the wheel. These objects are decorated with delicate stylised motifs, illustrating scenes from daily life as well as the first houses. In the 3rd century A.D., the expedition of Queen Jingu to Korea brought back ceramic models to Japan which influenced the decoration the great tombs, modelled on Chinese ones, in which decorated sarcophagi, bronze mirrors and "ha-

During the 18th century Buddhism spread to the countryside. Sculptures were realistic and didactic. The eyes of statues were often encrusted with rock-crystal (below). Japanese sculpture had finally freed itself from all Chinese influence (above).

niwa" (see under Miyazaki) have been found. These "haniwa", originally long terracotta cylinders serving to mark the entrance to a tumulus, were decorated with small incised figures of humans and animals. In fact, as Queen Jingu seems to have been a more or less legendary figure, it was due to Koreans who came and settled in Japan that these new, largely Chinese techniques became known here.

Buddhism arrives

In the 6th century Buddhism arrived in Japan, with its distinctive architecture and sculpture: the Buddha (Amida or Shaka), in the form of a man in his prime, meditating, with half-closed eyes and joined hands, or the right hand raised; Kannon, goddess of mercy; smiling, gracious Bodhisattvas (those shortly to become Buddha, and Rakans (disciples of the Buddha, their faces often distorted by passion). Under the protection of Prince Shotoku, Japan saw the Chinese architecture of the Horyuji temple take shape at Nara, with its five-storied pagoda and pillared buildings roofed with glazed tiles. The most ancient Buddhist sculpture is "Sakamuni triad", executed in gilded bronze in the year 623 (see description of this temple, under "Nara").

In 710 the Court established itself at Nara (see: *Japan through the centuries*). The new city was modelled on Tch'ang-an, the Chinese T'ang capital. The streets intersected at right angles and were lined with sacred buildings. The Imperial Palace was built facing south. Its wooden pillars are lacquered red and its roof is covered with green tiles. It is said to have taken twenty years, and a whole army of builders (as many as 50,000 carpenters and 370,000 labourers) to construct the vast Todai ji Temple to house a huge 17 metre high Buddha as well as hosts of priests and faithful. The Nara period was the golden age of sculpture in bronze, wood and lacquer. The faces portrayed are highly realistic.

Kyoto and "yamato"

In 794, the Emperor Kammu, fleeing from the influence of the Buddhist clergy, set up his court in a new capital, Heian (see: *Japan through the centuries*). The city (see: Kyoto) was again modelled on Chinese city layout, but at first the construction of monasteries was forbidden. A hundred years later, with the interruption of relations with the continent, art became detached from Chinese influences.

Yet Buddhism became so widespread that it influenced the arts afresh: paintings and sculptures abounded, reflecting the beliefs of the numerous new sects. The theme of "mandala" (heaven) became increasingly common. Temple architecture changed. The monasteries built on mountain tops (Enraku ji on Mount Hiei at Kyoto, Kongobu ji on Mount Koya) had to adapt their architecture to the configuration of the terrain. Usually they merged into the landscape.

The sects that rejected luxury and preached austerity instead simplified their temple buildings, seeking beauty in bareness and in the quality of the wood they used. They used wooden shingles for their roofs. Under the influence of the "jodo" sect, which preached faith in the mercy of Buddha and believed that to invoke his name was enough to gain entry to the "paradise of Amida", images became gentler and figures correspondingly much less austere. The architecture of this period is beautiful, calm and elegant. The temple of "Byodo-in" at Uji, near Kyoto (see: Kyoto), consists of a pavilion with two wings; reflected in its pool it seems almost like a bird about to take flight. The statue of Amida, by the sculptor Jocho, inside the pavilion, is made of wood, gilded and lacquered; it was made by putting together separate pieces which had been carved by different artists and then assembled in the Sanjo workshop in Kyoto.

When in 894 Japan closed her doors to China, "yamato" or "non-religious national art" developed. Its inspiration was Chinese painting, with its representations of the four seasons, famous sites, and agricultural scenes.

Painting came to express the Japanese artistic impulse at its highest, particularly in its sensitivity to nature.

Painters developed a capacity to express the changing moods of nature and of living beings. They drew their great skill from their proficiency at writing characters with brush and ink.

These paintings are of two kinds: "kakemono" is a painting on paper or on silk, which unrolls vertically; it is framed by an edging of woven material. When a painting is wound round a wooden cylinder, it becomes a picture-book to be read horizontally, known as "makimono".

In this national art the grandiose landscapes of China were replaced by the countryside and peaceful hills around Kyoto. These charming paintings illustrate stories—the twelfth-century Genji Monogatori among others. The scenes were first of all drawn in ink, then covered with gold and bright colours. These painters who were veritable miniaturists, also illustrated the lives of priests and the histories of the various temples.

Kamakura: the influence of "Zen"

When Minamoto Yoritomo settled at Kamakura (see: *Japan through the centuries* and Kamakura), he surrounded himself with warriors who spurned the aristocratic refinements of life in Kyoto. This new society was influenced once more by China and proved a fertile soil for a new religious sect, the "Zen" (see: *Religion*).

Their monasteries were very simple buildings and show a correspondingly basic rusticity in their treatment of materials. A typical example of this is the reliquary or "shariden", in the temple of Engaku ji, at Kamakura (see: Kamakura). Buddhism spread to the countryside, sculptures, designed to appeal to the uneducated masses, were often extremely realistic.

The eyes of the statues are often encrusted with crystal. Alongside the traditional school of Sanjo the new school of Unkei and his son Tankei, grew up.

Painting too was influenced by Buddhism and directed its appeal to the people. Painting completely replaced text on the scrolls, and religious beliefs are reflected in the gentle "paradise of Amida" and the tormented images of hell. The "Yamato" tradition continued, but the landscapes became less elegaic. In the illustrations to tales of war the incidents and figures are depicted in a very lively way. Portraits, such as that of Minamoto Yoritomo, accurately portray the characters of the "great men".

The Muromachi period

With their move into the Muromachi district (see: *Japan through the centuries*) the golden age of Ashikaga patronage began. They commissioned splendid palaces and elegant pavilions, like the Kinkagu ji and the Ginkaku ji, in Kyoto (see: Kyoto).

In such pavilions, world-weary aesthetes sought peace and lived lives of elegant refinement. The pavilion floors were covered with "tatami", and there were always flowers, arranged according to the rules of "ikebana", in the "tokonoma" (alcove), as well as a painted scroll.

Art entered into daily life. Every gesture was regulated according to canons of beauty. In special tea houses, built in the gardens, where everything was designed to rest both the spirit and the eye, a few cultivated people would gather to watch the tea ceremony (see: Chanoyu) and join for two or three hours in contemplation of beautiful gestures and beautiful objects. Precious bibelots were specially made for the "chanoyu".

The Momoyama period

After the exile of the Ashikaga shoguns, power fell into the energetic hands of Oda Nobunaga and Toyotomi Hideyoshi. Under them Japan was covered with castles: Osaka, Momoyama near Kyoto, and Hime ji, the finest of all (see "Hime ji"). These white fortresses with delicate roofs raised graceful forms above their battlements. They contained marvellous decorations by Kano Eitoku and his school.

An official painting academy

The Ashikaga had founded an official painting academy, where Chinese techniques were taught. With the development of the Kano school, artists depicted vast landscapes, taking in a whole ensemble in a few vigorous brush-strokes. Kano painters used predominantly black and white, adding a little grey, brown and gold. They painted vague ethereal landscapes, with sharp, jagged mountains, twisted pine trees, torrents and valleys.

The leading painter of this school was Sesshyu (15th century) who spent several years in China. His most famous picture, of Amano Hashidate, hangs in the Kyoto Museum. Kano Masanobu (15th century) painted misty landscapes of great refinement; Kano Eitoku was considered by the shogun to be the king of painters; his work unites ink monochrome with delicately coloured flowers and shimmering gold. Inside the castles of this period the walls and screens are often decorated with paintings by Eitoku, glowing gently in the diffused light from the "fusuma" (paper screens).

These paintings depict legends, the progress of the seasons, as well as popular and religious festivals. One of the loveliest works of this period is the "Screen of the Southern Barbarians", in Kobe (see: Kobe); it depicts the arrival of the missionaries in Kyushu in 1549.

The famous 17th-century painter Tanyu can also be considered a member of the Kano school; he decorated many rooms at Nikko. With a few deft brushstrokes he evokes a whole landscape. A picture of lions, painted in Indian ink, on a panel at Nikko, is considered to be his masterpiece.

The Edo period

Under the Tokugawa shoguns (see: *Japan through the centuries*) Japan closed her doors to foreigners. Internally it was a period of peace and order. There was a new rising class of city merchants and artisans. A few elegant palaces were built, retreats where intimates gathered for the "chanoyu" (places such as the Katsura Imperial Villa in Kyoto).

The first signs of bourgeois luxury appeared. Trading centres such as Osaka grew and prospered. Edo, the new capital, expanded daily. As the population grew, two-storey houses were built. There was always a great risk of fire. To counter it the use of plastered walls and tiled roofs was encouraged. Religious buildings and those erected for the shogun (at Nikko) were loaded with decorative ornament. When Buddhism fell from official favour decadence set in in sculpture and craftsmanship generally. Rival schools competed in every branch to supply goods of mediocre quality to the nouveaux riches.

Porcelain alone, an importation from Korea during the 16th century, retained a high artistic quality. The most sought after wares were those known as "Imari", produced at Arita, near Fukuoka. The greatest painter of the second half of the 17th century was called Korin; he was also noted for his fine lacquerwork. He is considered by the Japanese as one of their most typically national artists. His drawing is firm, powerful and flexible and his colour is delicate. He painted animals, flowers and humorous scenes. His compositions abound in fantasy and unexpected touches. He has been called an early impressionist...

*The landscape garden is seen as a work of art,
both a poem and a painting.
Each individual plant must be shown
to its best advantage and the whole
must be a harmonious,
non-geometric, rearrangement of nature.*

"Images of a floating, transient world".

Engraving techniques, imported from China, were probably practised in Japan as early as the 10th century. They were first used for Buddhist propaganda, to provide pious images of the goddess Kannon, for pilgrims to take away with them after visiting certain temples. In the 16th century Buddhist texts were illustrated with engravings; in the 18th century they appeared in the famous novels of the day.

The "Ukiyo e" or "Images of a floating, transient world" gave a new impulse to the Japanese print. "Ukiyo e" was a school of popular, realist genre painters; perhaps "actualist" might be a good term to describe them. They took people as their principal subject; at work, out walking, at home, at the baths. They paid great attention to settings; tea houses, theatres, gardens. They depicted places in a recognisable way. Some painters are known for their beautiful women with thick black hair, low foreheads, narrow eyes, long noses and narrow lips. The print was first of all considered as a souvenir to bring back from town, then it was used to illustrate almanacs and books. The four great masters of the art were: Utamaro (1752-1806), noted for his vivid pictures of places of amusement; Sharaku, who depicted the stylised mimicry of the "kabuki" actors; Hokusai, lively and realistic (the 36 views of Mount Fuji) and the landscape artist Hiroshige.

From the Meiji Era to the present day

The re-opening of Japan to the West, in 1868, released a passionate interest in everything Western. This enthusiasm in its turn led to a reaction. A Japanese Academy of Art was founded, under the direction of Okakura Kakuzo and the American Ernst Fenollosa, in order to protect and stimulate national art. Hishida Shunso (1875-1911) launched a revival of wash painting. The Fine Arts Academy was established, to perpetuate the traditions of the Kano and Korin school.

Today, Kaii Higashiyama is the leading Japanese traditional painter. Born in 1918, he took courses at the Tokyo Fine Arts School and then studied in Europe. In 1947 he won first prize at the "Expo Nitten" and since then has gone from success to success. He has been commissioned to do mural paintings in the Imperial Palace and in the Togu Palace, residence of the Crown Prince. He has also painted "fusuma" (sliding screens) for the ancient temple of Toshodai ji, at Nara.

His vast painting "The Sea at Dawn", which he did for the antichamber of the Imperial Palace, shows two black rocks rising from the trough of a fascinating sea-green wave, flecked with foam; this picture symbolises Japan, surrounded by oceans. Among his best known works we might mention: "Evening Calm", "Peaceful Autumn"—an extraordinary cone of russet brown foliage seen against a grey sky, and "Clouds over the Mountains" —bluish tones evoking trails of mist among pine trees on the hilltops.

The Japanese garden

Her gardens are among Japan's loveliest artistic treasures. They are all—whether pocket handkerchiefs or parks—conceived of as works of art. And how the Japanese love them! They visit them in crowds, particularly in blossom time. Gardens seem an appropriate expression, of the feeling for nature that is such an important part of Shinto belief. Buddhism too, preaching as it does the brotherhood of all living things, men, animals and plants, only reinforces this.

There are magnificent parks (Ritsurin, at Takamatsu and Korakuen, at Okayama) and lovely villa gardens (Katsura, at Kyoto). Yet just as important is the square metre of earth that the (fortunate) Tokyo citydweller has in front of his house, and tends with all the care of a landscape architect, anxious to bring out the beauty of every single plant. Even a bowl of soil, with three tiny ferns and two miniature shrubs, is enough to give him scope for a perfect expression of natural beauty.

The art of the Japanese garden dates from the earliest days of Kyoto in the 9th century, during the cultured Heian period. It was forgotten for a while, during the war-torn Middle Ages, but revived and reached its peak during the 15th and 16th centuries.

The early gardens were inspired by those of T'ang China (618-906). A lake, or a lotus-covered pool, rivulets, waterfalls and delicate bridges, islands and pavilions for the tea ceremony—these were all essential elements. The life of the Court and of the aristocracy often had such a garden setting: "gay, elegant, open and bathed in sunshine".

The landscape garden (those of the villa Katsura and the Heian Shrine, for example) was designed so that guests who attended the tea ceremony there would pass through it, forgetting their cares and relax into the mood of calm self-control appropriate to the ceremony itself. There was nothing geometric about its design. Trees, shrubs and flowers were carefully arranged so as to form a harmonious re-creation of nature.

Three essential constituents, stone, trees and water, were used to compose a living whole. Stone, to symbolise the bone structure, was used in various ways—rocks, stone lanterns and flagged areas and pathways; trees (hair) and water (blood), made up the trio. Fallen leaves and moss induced a mood of meditative calm. All these elements were ordered in accordance with a variety of refined aesthetic doctrines. One of these insisted that shrubs be placed in the foreground, with trees and hills behind, to symbolise the immensity of nature (e.g. the Shugaku Villa at Kyoto). Another, wishing to emphasise distance, placed hills and large trees in the foreground. Pines, cedars, cryptomerias, smaller trees such as maples and cherries, laurels and azealeas—chosen for the colour of their flowers and leaves—as well as dwarf trees, were all used. Flowering plants were never arranged in geometrical beds.

Winding stone-flagged paths, designed to lead the visitor to the important viewpoints, were important aids to appreciation and to meditation. (Though more down-to-earth observers claim they were mainly laid to prevent visitors walking in the mud and dirtying their kimonos.) Many gardens were modelled on famous prototypes or natural features in China or elsewehere in Japan, as at Matsushima and Amano Hashidate, or the gardens of Koraku en, at Okayama, with their large lakes and islands which symbolise the Inland Sea.

The gardens conceived according to Buddhist ideas are completely different. When they arrived in Japan during the 14th century, the Zen priests, seeing nature as a personification of the Buddha, gave a philosophical basis to garden design. Enlightenment being achieved by intuition, the garden was seen as a link with the mysteries of life "a bridge between being and non-being". Such gardens exclude any allusion to violence. They attempt to arrest nature, avoiding changes from season to season and making use of deliberately stunted trees that will not grow or change their shape. The whole garden has to reflect the immutability of nature. Any tree with the irritating habit of growing bigger or losing its leaves is rigorously excluded. Such rigour inspired gardens composed of sand and stones. They breath the calm of eternity, offering the viewer great depths and infinite horizons. Their stones are charged with different meaning according to their shapes, their texture and the angles at which they stand.

Typical Zen gardens are those of Ryoan ji and Nishi Hongan ji, at Kyoto. To a Western observer they are both fascinating and baffling.

"Landscape" gardens, on the other hand, are a tourist's delight. The visitor to Japan can see an enormous variety of them—great

*The freshly-dyed silks,
future obis and kimonos,
are spread out to dry on the river bank.
In the Nishi-jin quarter, at Kyoto,
a centre of the silk industry.*

parks, villa gardens, country gardens embowering wooden houses, narrow borders of green to set off a provincial facade; the hidden gardens of the "O chaya" (tea houses) and, last but not least, the exhibition of miniature gardens in some little railway hall, with station masters and their assistants bending down to gaze at them with tender solicitude!

Ikebana

The art of flower arrangement was an importation from China, along with Buddhism, whose temples' altars were lavishly decked with flowers. It found an early patron in the person of Prince Shotoku who in the 7th century it is said, taught the art to Ono no Imoko, founder of the first school of ikebana. During the 15th century, when the Gold and Silver Pavilions and the Tea Pavilions were built, the art reached full public recognition. It became customary to place in the alcove, "tokonoma", an "ikebana" or arrangement, to harmonise with the "kakemono" or painted scroll, which hung there. Gradually rules were worked out to ensure that the arrangements harmonised perfectly with the painted scroll, the season, the occasion (wedding, festival) and the character of the guests.

Whereas in the West flowers are usually bunched together for colourful effect, in Japan each flower is treated individually. Instead of being thrust together in a vase each bloom must be carefully placed so that its individual beauty can be appreciated. An arrangement is a matter of putting together flowering elements of different shapes and sizes to form a picture of pleasing lines and forms. Quantity must not be allowed to stifle the beauty of the single flower. The arrangement may be symbolic or express some philosophical idea. Symetry must be avoided at all costs. Flowers must be arranged naturally, respecting the way in which they grow. An irregular triangle may be formed of sprays of different lengths; the tallest stands for heaven, the next

Every craftsman understands and loves his medium; from this communion stems the beauty of the objects he creates.

for man and the shortest for the earth. The whole arrangement must harmonise with the kakemono; thus in the autumn, with a kakemono depicting brown-leaved maples, it would be suitable to place an arrangement of green-leaved autumn plants; with a mountain landscape, alpine flowers would be appropriate, and so on.

Flowers in season are always to be preferred; so, in the spring, cherry-blossom is appropriate; in May, wisteria; in the autumn, chrysanthemums.

The choice of vase is significant. Reeds and irises arranged in a pair of vases express simplicity and calm. Passion is symbolised by a pine branch standing in a bronze container... Sometimes a message may be conveyed by a flower arrangement —thus a "farewell" arrangement must contain some flowers symbolising return...

During the last few years the art of ikebana has been so enthusiastically taken up by people in the West that it is perhaps unnecessary to say more about it here. However, any visitor to Japan who is invited to a Japanese home and wishes to take flowers to his hosts is strongly advised to mention this to the florist; who will carefully compose a bouquet with this in mind—the hosts in turn will be delighted and flattered at this mark of consideration.

The mystery of lacquerwork

Earlier than the 6th century lacquer was used in Japan, on wickerwork, combs and bows. From the 7th century on, lacquerwork was extensively used in temples and shrines. In the Nara period (8th cent.) the government encouraged the production of lacquer by giving each family a piece of land to be reserved for lacquer-trees. It seems likely that lacquer boxes, designed to contain sacred writings, were made by the monk Kobodaishi. Under the Tokugawa's (1603-1868) the fashion for lacquerwork spread and it was used to cover doors, and columns in temples and palaces, as well as small objectifs for everyday use such as small boxes and trays.

Lacquerwork is a highly skilled craft. The lacquer itself, the sap of the lacquer-tree, Rhus vernicifera (urushi-no-ki, in Japanese), is used either pure or mixed with powdered pumice-stone or iron sulphate. The object to be lacquered is carefully carved out of wood and stuck together if necessary; it is then rubbed down with finest pumice-stone. A first layer of lacquer is then applied, then a layer of Indian ink, then another layer of lacquer. The object is then set to dry in a dark, slightly damp place. Then more layers of lacquer are applied. Lacquerwork can be black, gold (green, red or yellow), red and green. It can be decorated with mother of pearl, ivory or metal, as well as with incised designs.

The art of present giving

The Japanese must find the Western style of present giving quite barbaric! Japan is the country of presents, par excellence; they give rise to a whole industry. People take enormous pleasure in giving and receiving presents. Any occasion will serve as an excuse: festival, visit, birth, marriage, death, arrival, departure, examination... Like the "language of flowers" there is an etiquette for present giving: dolls are given during the children's festival; aged relatives receive birthday presents of cakes and warm clothes; rice and fruit are the usual offerings at Buddhist services, money at Shinto ones (see also under Religion).

Of course, the value of the present varies according to the position of the donor and the recipient. It would be a breach of good manners to respond to a present by giving one of greater value in return. The recipient would be embarrassed and feel obliged to reply with an even more valuable gift. At this rate, both sides would soon be ruined! Money, tactfully offered in an envelope, is always a welcome present...

The wrapping of a present is an art in itself. The merest trifle, bought in the foyer of the Kabuki theatre or at the entrance to a shrine or temple, even a handful of sweets bought from the street-vendor, will all be swiftly and beautifully wrapped for presentation.

The Western visitor will soon be struck by the number of little shops to be found at all the sights, near the temples, and in the station halls in the tourist resorts. A Japanese woman merely dining at a restaurant with a friend will leave with two or three packets of sweets in her hand —presents for her family (note the number of little parcels already prepared, waiting by the cash desk); a journey, however short, will involve numerous presents. It would seem rude to return from a week-end away without a little present for each member the family and for one's friends... So, at places like Amano Hashidate and Takayama, the streets are lined with shops selling souvenir-presents and station platforms are crowded with returning visitors, all laden with parcels, which they often carry in the traditional cotton square, the "furoshiki", tied together at the corners.

For the Japanese, one of the main pleasures of travel is without a doubt the purchase of such presents; Japanese etiquette and tradition are much encouraged in this by the "present industry."

What do the Japanese give each other? Presents range from boxes of biscuits, beautifully packed (the boxes bear pictures of the local sights—Mount Aso in Kyushu, a bear in Hokkaido), to boxes of dried fish or dried sea-weed from the seaside, to dried octopus from Amano Hashidate, or carved wooden bears from Shiraoi.

It will be much appreciated if the Western visitor takes a present for his Japanese friends, a trifle from London or New York, that he has brought with him, or souvenirs of his travels in Japan itself. □

THE STONE GARDEN OF RYOAN JI

■ *A very old monk carefully swept his garden, as he did every morning, and then went into the shrine to meditate. Seeing a novice at his prayers, and wishing to test him, he tapped him gently on the shoulder and said: "Go and tidy the garden." The young monk went out and saw that the sand had been beautifully raked, the moss had been brushed and every tree pruned just as it should be. Not knowing what to do he returned to the shrine, where the old priest watched him with wry amusement. Then he suddenly stood up for he had noticed a young maple whose golden autumn leaves were the only splash of colour against the moss and sand. He struck the tree a light blow with his hand; a leaf floated slowly down onto the ground. Having done this, the young monk returned to his prayers. "Quite right", said the old monk.*

ELISSEEFF
Légende sur le jardin Zen

literature and the theatre

■ Over the centuries Japanese literature illustrates and gives depth to the history of the country. It expresses certain important features of the Japanese national character: respect for the past, love of nature and family, patriotism, courage and gaiety. At first it was influenced by China from whom Japanese writing took its ideograms. At the beginning there were two sorts of written works: those of "men of letters", cultivated people influenced by Chinese thought, and those of humbler writers who did not take happily to these foreign influences.

Literature

The first literary works were simple songs and prayers addressed to Shinto gods. In the 8th century, during the Nara period, the first great prose work appeared, the "Kojiki", a sort of Bible of Shintoism which recounts its myths and legends. At this time too Japanese poetry began to flourish. It was unrhymed and consisted of alternating five and seven syllable lines. Lafcadio Hearn said of it: "In Japan this poetry was as universal as the air, it was heard by all, read by all and composed by almost every one, without distinction of class or social condition."

The Japanese are less sensitive to the music of a poetic line than to the picture conjured up by the ideograms. From the 8th century, Japanese poets have excelled in the composition of very short poems. The "Tanka" is a poem of five lines which contain thirty one syllables in all.

The usual themes are impressions of nature, love, sensations, feelings expressed in a highly condensed form. The poet uses "pillow words" (Makurakotoba), conventional epithets, and "pivot words" (Kakekotoba) which can be used in two different ways, one referring to what precedes and the other to what follows. A seventh-century anthology, the "Manyoshu", consists of some 4,000 such poems.

"The sky is a sea
where the clouds rise up like waves;
the moon is a boat
being rowed towards
the thickets of stars, as if to hide there."

Kakinomoto Hitomaro

The Heian Period (9th to 12th centuries) has been called the "classical age" of Japanese literature. Chinese influence ceased after 894, when a "closed door" policy was adopted with regard to China. Both the written and spoken language gradually changed, as did the Japanese conception of poetry. It became less philosophical, in the Chinese manner, and more personal, more concerned with formal structure and detail.

At this time there was a great vogue for personal journals and for narrative tales ("monogatori"). Women occupied a privileged position in this sophisticated society and were responsible for two of its literary masterpieces. Murasaki Shikibu, a lady of the Court, wrote the "Genji Monogatari" ("The Tale of Genji"), around the year 1000; it tells of the amorous adventures of a prince, against a setting of fashionable Court life.

Several chapters of this work, which is rated by many as the supreme masterpiece of Japanese literature, were translated into English, as long ago as 1882, by Suyematsu Kenchio. The "Makura no Soshi" ("notes of the pillow"), is a collection of thoughts, philosophical reflections, and subtle and amusing sketches of Court life.

The years between the 13th and the 16th centuries were a period of literary decadence. It was a time of incessant warfare and culture was neglected. There was some interest in historical tales, and in epic stories recounted in rythmical prose; their themes were the wars themselves, the fall of the Fujiwara and the setting up of the Kamakura regime. (See: *Japan through the centuries*.) The "Heike Monogatari" tells of the

*Noh masks are carved out of wood,
then painted and lacquered.
They express not so much the feelings
of their characters but their essential
nature—god, demon, spirit, old man, warrior;
this mask represents a blind man.*

struggle between the Minamoto and the Taira families.

A few educated people interested themselves in Buddhism, which was spreading during the period to all classes of society. In the "Hojoki", or "Life of a Hut Ten Feet Square", a courtier tells of the misadventures that have brought him to a Buddhist detachment, and talks of the transitoriness of things: "A river flows without ceasing, but the water is never the same; the foam that floats and stops before an eddy, disappears then reappears, but it never lasts for long. Such is the life of man and his dwellings in this world..."

At the end of the 14th century, a new literary form appeared, the Noh. In the era of peace that began during the 17th century, a very different type of literature arose (short poems, popular plays and novels), appealing to the people, the commercial middle classes, and to the "samurai" who (their masters having been overthrown by the Tokugawas) took to teaching poetry and painting. A new kind of witty poetry became fashionable, often modelled on the "haiku", in which a minimum of words evoking landscape, sounds and smells, were used to make a vivid impression on the reader.

"If you put a handle on the moon, what a fine mirror, it would make!"

Yamazaki Sokan

The popular theatre, the "kabuki", made its appearance with the works of Chikamatsu Monzaemon and Takeda Izumo. These two dramatists produced the famous play "The 47 Ronins" (see special entry).

The comic novel was a popular form at this time too. The most famous of them is "Hizakurige", by Ikku (1795-1831); in it he relates, with Rabelaisian verve, the adventures—all over Japan—of a pair of comic, good humoured rakes. There is a nice story told of Ikku himself: he asked to be cremated in the clothes in

TRADITIONAL JAPANESE MUSIC

■ *The ancient Shinto dances (see: Religion) used to be accompanied by primitive music from flutes and drums. Japanese music proper grew out of Buddhist ritual. At first intended to give backing to religious chant, it was soon used to accompany the chorus in "Noh", in "Kabuki" plays, recitations of passages from novels, and the songs and dances of the geishas. Court music was known as "gagaku"; popular music as "zokkyoku". The "biwa", an instrument of Chinese origin, introduced into Japan during the 10th century, is a sort of flat mandoline with four strings made of silk impregnated with lacquer. The "koto", another Chinese importation, much esteemed by connoisseurs, looks like a large horizontal harp with thirteen strings. To play it, the musician fixes ivory "fingernails" to his fingers and kneels down facing the instrument. The "samisen", a three-stringed guitar, imported from Ryuku (Okinawa), is the geishas' favourite instrument. The "kokyu", a small two or three-stringed violin is often played in trios with the "samisen" and the "koto". The samurai used to play the "shakuhachi", a bamboo flute. The "fuye" is a seven-hole flute, and the "tsuzumi" is a sort of tamburine, shaped like an hour-glass.*

which he had died, and then astonished his assembled mourners with a succession explosions from firecrackers he had concealed about his person!

The Meiji Restoration brought contacts with Europe. Until then, the only European literature known in Japan was Aesop's "Fables", translated by a 16th-century missionary, and "Gulliver's Travels". A whole range of European dramatists and novelists were quickly translated: Cervantes, Fenelon, Shakespeare... The establishment of a Japanese press, in 1868, brought great changes; within five years there were a hundred periodicals in circulation. A new genre was created, political literature. Numerous foreigners were invited to visit Japan, bringing new ideas.

"The friends of the Ink-Stone"

Some Japanese were so carried away by their enthusiasm for the West that they rejected their own national literature, whilst others, anxious to preserve a pure Japanese style, founded a movement known as "The Friends of the Ink-Stone". Literary realism, introduced by Tsubouchi Shoyo (1839-1933), gradually gave way to social concern. At the beginning of this century, Japanese writers inclined to a sort of naturalism, inspired by Emile Zola; at this time too, socialist ideas began to affect literature. Disillusioned, certain writers sought refuge in history. One of them, Akutagawa Ryunosuke, is best known as the author of "The Gate of Rasho" or "Rashomon".

After the collapse of security and belief, following the Second World War, there was an upsurge of pacifist ideas. At the same time a "literature of the flesh" made its appearance...

Among contemporary writers, Kawabata Yasunari, who won the Nobel Prize in 1968 and committed suicide in 1972, is well-known in Europe. His novels, translated into English as "Snow Country" and "Thousand Cranes" are notable for their great sensitivity and delicacy.

Yukio Mishima (1925-1970) was a talented writer who wrote in a neo-classical style. His work is a successful blend of Western and Eastern culture. In "Confessions of a Mask" he recounts his early life; "After the Banquet" describes the intrigues involved in the election of a Governor of Tokyo; "The Golden Pavilion" reveals the writer's personality, his aestheticism and his acute psychological perception. At the age of forty-five Mishima, who all his life had been fascinated by death, committed ritual suicide, "seppuku". Just before he died he had sent his publisher his last work, four stylish novels with the collective title "The Sea of Fertility." Mishima will remain an important figure in world literature.

Endo Shusaku, born in 1923, is that rarity in Japan, a Catholic writer. His continual preoccupation is the conflict between traditional culture and faith and religion. His best known works are "The Silence" and "The Dead Sea". All these modern novels exist in English translations.

The "kabuki" theatre

A six-hour visit to the "kabuki" theatre is nothing out of the ordinary to the Japanese; they do not come to see just a single play, rather a whole day's programme of four or five, with interludes of mime and dance.

The audience of city-dwellers and people from the provinces and the countryside arrives around eleven o'clock in the morning; they make themselves thoroughly at home in the theatre, and frequently stay until the lights go down in the evening. They often move about, applauding their favourite plays, and skipping the scenes they find boring to join their friends in the foyer, where there are comfortable seats for rest and conversation, as well as little restaurants and stalls selling drinks and souvenirs.

The origins of the kabuki theatre are somewhat obscure. It possibly stems from those festivities where women recited passages from epic tales, to the music of the "shamisen",

*Elaborate production, gorgeous costumes,
splendid scenery, contained violence
in speech and acting,
the popular "Kabuki" theatre is at one and the same time
classical theatre, melodrama, opera and music hall.*

as in the "kyogen" or interludes of the Noh.

The Kabuki dance is said to have been created, at the beginning of the 17th century, by a woman named Okuni. She was a priestess in a temple and a great beauty. She fell in love with a "ronin" (a masterless Samurai warrior), Nagoya Sanzaburo, and fled with him to Kyoto. It was the custom there, during the summer, to present Buddhist dances on a stage erected in the dry bed of the river Kamogawa. Okuni is said to have danced there to earn a living and then moved on to Edo (Tokyo) with her lover, where she recruited actors and founded a theatre.

In 1667, following strife among the actors, an edict forbade men and women from appearing together on the stage. From then on women's parts were played by men. Another edict, in 1681, forbidding the wearing of swords in the theatre, had the effect of excluding the samurai, who refused to relinquish their weapons. The nobility abondoned the theatre, which thus became the pastime of the commercial and lower classes. By attending the theatre himself, in 1887, the Emperor Meiji dramatically rehabilitated the Kabuki.

Kabuki: ever alive

From its very beginnings, two differents sorts of plays were performed in the Kabuki theatre: plays about everyday life and historical dramas. Today most plays by contemporary authors are set either in feudal times or during the Meiji period. The play that really draws in the crowds, and which is often revived when audiences tire of modern dramas, is "The Adventure of the 47 Ronins" (see special entry). It has been said that, for the Japanese, the Kabuki is at the same time their classical theatre, their opera and their music hall!

Kabuki is performed in large modern buildings. Behind a classic Japanese facade the Tokyo Theatre consists of a huge European-style auditorium. A revolving stage and other complicated machinery makes it possible to mount elaborate productions with rapid changes of scene. The scenery itself is often the work of distinguished artists. The "flowery path", a sort of gangway, runs across the front of the house; hidden passages link it with the wings. The actors wear wigs and are heavily made up; they wear elaborate costumes. The plays unfold extremely slowly, the actors' voices are charged with contained violence. "A Kabuki play is a succession of tableaux in which the muscular tension of the actors emphasises and prolongs the impact of the text itself" (Elisseeff). The orchestra, composed of traditional instruments, further underlines the action and the atmosphere. Moments of great pathos are indicated by the knocking of two wooden blocks on a board on the floor.

The Kabuki remains a popular entertainment, young and old flock to the theatre during the season. The foyer is alive with pretty girls in light-coloured kimonos and older women wearing carefully chosen "obi" (wide sashes), walking delicately on their "geta". Everyone takes great delight in these medieval tragi-comedies, with their lengthy monologues and their action often as slow as life itself. The audience keenly awaits the first appearance of the hero on the "flowery path", buzzes with appreciation for a perfectly-delivered speech and will warmly applaud an actor's faultless piece of by-play with a sunshade (it may last ten minutes!); they cheer the hero and hiss the villain... and never cease to admire the scenery... The visitor, clutching his (English) programme may well find himself muttering: "Incredible... magnificent... but, really, too long... and how odd, all these men dressed as women"—only to realise that he too has been held spellbound for six hours by the Kabuki theatre.

"Forget the theatre and see the Noh; forget the Noh and see the actor; forget the actor and see the idea, forget the idea and you will understand Noh." (Zeami).

For the uninstructed foreigner a Noh play is strange experience, vastly different from anything he may have encountered in a Western theatre. There are descriptions, in 8th century religious texts, of formalised dance mimes derived from the dances of Ameno Uzume when she tried to lure the goddess Amaterasu from the cave where she had hidden (see: "*Religion*"). During the 14th century these religious song and dance mimes were further animated by dialogue, inspired by the recitation of passages from famous novels such as the "Heike Monogatori".

The Noh

Groups of travelling players were formed, who performed at festivities organised by the nobility. Others, known as "Za", settled down near temples, performed more elaborate dramas and often became prosperous. Early in the 15th century the shogun Yoshimasa became interested in the art and, following his example, later shoguns had their own appointed companies—in which their "daimyo" (vassals) sometimes performed as well.

Most Noh plays date from the 15th century. They were often written by Buddhist monks. The actor Zeami Motokiyo was responsible for more than a hundred Noh plays and for interpretive works as well, in which he tried to define and perfect the notion of "yugen"—a quest for "profound and mysterious" reality. After his death different schools of theatre developed. The Kabuki was essentially a popular art, whilst the Noh appealed particularly to the nobility and to the warrior class. There was a temporary falling-off during the Meiji period but the Noh soon recovered and remains a highly-esteemed form of theatre in present-day Japan.

Noh plays used to be performed in the open air, often in temple courtyards. The most ancient Noh stage is still in existence, in the Itsukushima temple, near Hiroshima. Today Noh plays are sometimes performed in the traditional manner during religious festivals, but they are usually given in relatively small theatres. The audience wear kimonos—jeans and sweaters would be totally out of place at a Noh play—and sit on cushions on the floor which is covered with tatami (woven mats); everyone has a copy of the text of the play. The stage is a square structure, five metres across, with a Buddhist-style roof supported on four pillars; at the back there is a rectangular space for the musicians; on the left a "bridge" links the wings to the stage proper—on one side of it there is a balustrade and three little pine trees and at the other end there is a sort of tent. The stage is surrounded by a gravelled area which recalls the temple courtyards...

In complete silence, the musicians enter one by one and take up their appointed places, the flautist first, then a man with a small tambourine, then another with a big one; if the hero of the play is a god or a devil there will be a big drum as well. Then the chorus enters. Finally the scenery is brought on, highly stylised and symbolic: a dwarf tree to represent a forest, a piece of wood to represent a ship.

The Noh repertoire at the moment consists of two hundred and forty plays, grouped according to subject: ancient legends, moral tales, tales of gods and spirits, and of ghosts seeking vengeance. A Noh performance always consists of five plays of different genres, each lasting at most an hour. They are written in archaic prose interspersed with lyrical passages, all in sentences of five and seven syllabes. There is frequent and painstaking use of "pivot" words: the last word of one sentence and the beginning of the next is often the same, but used in a different sense.

There are only two actors in Noh: the "waki" a passive attendant character who provokes the "shite", the active character, to speak. "The 'shite' is the ambassador of the unknown... a secret, hidden being that comes to seek enlightenment from

THE ADVENTURE OF THE 47 RONIN

■ *In the year 1701, in the Shogun's palace, Lord Kira insulted Lord Asano. Asano drew his sword and sprang at Kira who fled. Unable to avenge his honour Asano had no alternative but to commit "seppuku" (suicide by the sword). His vassals thus became "ronin" or wandering, masterless "samurai". They swore to avenge their lord. They spent two years waiting for their moment. Eventually, one snowy night, they entered Kira's castle and cut off his head. They carefully washed the severed head, for an inferior must appear clean before his superior and Kira had become Asano's inferior. Then they took the head, together with the sword that had been used, and a letter recounting what they had done, and placed them all on Asano's grave. Finally, no longer having any aim in life and to avoid punishment for their "crime", all 47 Ronin committed "seppuku" and were buried beside their lord. This Kabuki plot is based on a true story. The graves of the 47 Ronin are in the Temple of Sengaku ji, in Tokyo (see: Tokyo).*

the 'waki'. The 'waki' questions, the 'shite' replies, the chorus comments and creates a musical setting of words and images that enfolds the mysterious masked pathetic figures sprung from the void." (Paul Claudel).

Noh actors: symbolic figures

Noh actors are essentially symbolic figures. They wear magnificent costumes and masks bearing sweet, sad or tragic expressions. The mask denotes both the nature of the character and the mood of the play. The gestures of the actors are highly stylised: two gliding paces forwards to indicate joy, a single step backwards to show sorrow. Thus the text, spoken in a harsh impersonal tone, is enveloped in a sort of slow dance. The orchestral contribution to the unreal, poignant atmosphere is often very noisy—shouts, howls and plaintive cries.

Every Noh play evokes an atmosphere of extreme tension in the theatre: "The object of the Noh is to enable the audience to enter into an ideal communion with a higher reality to which costume, scenery, music, words and song all contribute, though it lies above and beyond them all..." (Louis Frederic).

The audience put their texts down, gripped by an almost desperate expectancy that the "shite" will grant them at least a glimpse of a world beyond: "With one wave of his magic fan the 'shite' sweeps away the present as if it were a mist and commands the past to rise up around him." (Paul Claudel).

The Noh is quite a test of nerves for those who watch it. It is customary to play a comic scene between the plays in order to give the audience a chance to relax. These farce interludes, known as "Kyogen" are rather similar in their coarse humour to medieval French Fabliaux.

The foreign visitor may well find Noh very difficult to take, at first, but he will very probably be caught up by it after a while and experience-

78 PANORAMA

*Japan is a land of festivals.
The Gion Matsuri, on 17th July,
stems from an 11th century procession.
Enormous floats of gilded wood, held to be
the earthly dwellings of the Shinto gods,
are escorted through the streets by priests,
musicians and men on horseback.*

—like the Japanese—an almost physical paralysis as his mind is drawn to its incommunicable mystery. Dazed by the experience he may leave the theatre doubting whether he will ever understand the soul of Japan as is portrayed by the Noh, a drama reduced to essentials, with its appeal to mystery and nothingness. Even today the Kabuki theatre attracts a wide cross-section of Japanese society whilst the Noh remains the more private delight of the intellectual elite.

The Bunraku za or puppet theatre

The first puppets came from China, probably around the 10th century. Itinerant beggars are said to have used them in little shows in country villages. Some performed in temple courtyards. Various folk-tales were added to these puppet plays, known as "Joruri"—after a heroine of the time when the Heike and Taira clans were locked in combat (see: *Japan through the centuries*). The first "Bunraku za" was founded in Osaka at the end of the 18th century; soon every city of any importance had its own puppet school. The great contemporary dramatist, Chikamatsu Monzaemon, wrote dramas and historical plays for these puppet theatres.

When the curtain rises, the play is announced by a narrator, who then begins to tell the story. The puppets enter. They are large, 1 metre 30 centimetres tall, and sumptuously dressed. They are not worked by wires but each has three operators, dressed in black. Instead of a body the puppet has a bamboo framework to support its costume. Their wooden heads have real hair and they are held up by a rod in which there are three levers, to move the neck, the eyebrows and the mouth. Their hands are movable as well. The principal manipulator—the only one to appear barefaced—the others wear masks—operates the levers that move the head, through a hole in the back. He manipulates the puppet's right hand with his own right hand; the second man operates the puppet's left hand, the third his feet. They manage to move their puppet with perfect synchronisation; it really seems to come alive, its gestures are natural, it opens and closes its eyes and mouth in a thoroughly realistic way. "The moving doll takes on the collective soul of this shadowy stage..." (Paul Claudel).

The Bunraku has a highly dramatic repertoire. The plays are long —they can last up to twelve hours. Shows tend to consist, therefore, of extracts taken from different plays. For sheer literary quality Bunraku can be said to be the true classical theatre of Japan. Every large city has its puppet theatre, but they do not play all the year round.

The Japanese cinema

The Japanese cinema is going through a crisis at present. In 1960, when the post-war cinema boom was at its peak, there were 7,400 cinemas in Japan. In recent years two thirds of them have been converted into offices or car-parks. Film production, which reached an annual average of seven hundred feature films in the palmy days, fell to forty in 1977. Out of one hundred Japanese filmgoers in 1960, only ten were still faithful in 1977. Faced with American competition and judging its own films too difficult for foreigners to understand, Japan has abandoned her attempts to export them.

Television, with its many channels, is mainly responsible for this parlous situation. Some directors, like Kinoshita, have "recycled" themselves for television, where they make films and above all, serials. The latter are extremely popular, comprising thirty to fifty one-hour episodes. Every fifteen minutes, serial broadcasts are interrupted for three minutes' advertising on behalf of the firms—usually at least four—who have sponsored the film.

According to statistics, a serial is watched by some seven and a half million viewers. Films, often erotic, are still being shown on some channels at midnight.

Half of the current production of forty films a year are erotic, "nedoko", and pornographic, the latter often strongly tinged with sado-masochism. The rest are realist films, films of social criticism, films with a medieval setting, and comedies.

Japanese films are comparatively unknown in the West. One may sometimes be shown in London, Paris or New York (often a film that has won a prize in a festival years previously—like "Red Beard" which came out in 1965 and finally appeared in Western cinemas in 1978) but provincial centres seldom see them. Often Japanese films are somewhat confusing, due to their medieval settings or their frequent use of traditional and symbolic references.

The great directors

Yet although the Japanese cinema is going through a difficult time at the moment, it has had its hours of glory. Directors such as Kenji Mizoguchi, Teinosuke Kinugasa, Yasujir Ozu, Akira Kurosawa, Kobayashi Masaki, Kon Ichikawa, Kaneto Shindo, Suji Teraryama, Oshima, and Masahiro Shinoda are considered by the Japanese to be the true masters of this seventh art.

Kenji Mizoguchi achieved fame slowly. Among his early films were "Naniwa Elegy" and "Sisters of Gion", they appeared in 1936. He was concerned with the position of women in society and strongly critical of the tyranny exercised by family and tradition. "The Life of O'Haru" (1952) gives a picture of the sad life of a beautiful woman who becomes old and ugly before she dies. His "Ugetsu monogatari" (1953), stories in which realistic and irrational elements were blended, made him better known in Europe. His last film, "Street of Shame" (1956) was acclaimed at the Venice film festival; it is a realist film set in a modern brothel.

Teinosuke Kinugasa (born in 1898) has campaigned for the recognition of the autonomy of the cinematic art. "A Crazy Page" (1926), a silent film, put him at the forefront of the avant-garde. In 1971 he came across a lost reel of this film and re issued it, with a sound-track. The film, the fruit of a collaboration between Kinugasa and the novelist Kawabata, tells the story of a sailor who takes a job in a mental hospital in which his wife is a patient. Kinugasa visited Europe between the wars and studied under Eisenstein. His most famous films are: "Before Dawn" (1931), "The Loyal 47 Ronin" (1932), and "Gate of Hell" (Cannes Grand Prix, 1954).

The work of Yasujiro Ozu (1903-1963), belongs to the "Shimun Geki" school; he made human dramas and comedies with working or lower middle-class heroes. "Tokyo Story", which won a prize in London in 1953, is reckoned to be his masterpiece.

One of the best living directors is Akira Kurosawa (born 1910). "Rashomon", which won the Golden Lion at Venice in 1961, was a revelation to Western cinema audiences who had previously thought that the Japanese only made films about samurai. As early as 1946, he portrayed the sufferings of post-war Japan in his film "Drunken Angel", followed by "Quiet Duel", and "Stray Dog". "Red Beard" (1965) is "the code of conduct of the wise man, Akahige (Red Beard), designed to produce superior men, noble beings, cultivated, intelligent and disciplined gentlemen" (Kochi Yamada).

Among the contemporary directors who are trying to breathe new life into the Japanese cinema, mention must be made of Kaneto Shindo. His finest film is "The Island" (1960), the story of a peasant family and its struggle against the elements and the seasons. It is based on Shindo's own childhool: "I wanted to show how the peasants struggle with the earth. My Mother never spoke of her silent battle with the forces of nature. This impressed me, so I thought of a film without dialogue, in which the pictures would be the only narrators."

In "Harakiri" (1962) Nasaki Kobayashi (born 1916) portrays authority, the military hierarchy of the 18th century. "I use history to depict the present. I think of this as a highly contemporary film."

Masahiro Shinoda based his film "Sapporo" on the Olympic Games of 1972. The poet-producer Suji Teraryama experiments in his films with what he calls "the fictionalisation of the image". His bold, baroque film: "Lets throw away our books" (1971) is a story of the crisis of an adolescent, of his relations with his family and his first sexual experiences.

Television in Japan

Television has certainly had a dramatic effect on the Japanese cinema as it was ten years ago. But yet it is itself, in some sense, its modern continuation. Audiences of ten million now see the classics of Japanese film-making, albeit on a smaller screen; and television continues to provide interesting openings for film directors who have "recycled" themselves to cope with its challenges. Thus television makes for the survival of the cinema, if not for its renewal. ☐

*The national Sumo competition,
"the honourable struggle" stemming from ancient traditional wrestling
(during the 18th century bouts used to
be held in shrine courtyards),
takes place six times a year, in Tokyo.*

PANORAMA 83

the japanese table

■ The arts of cooking and the table are practised, in Japan, with the utmost refinement. A Japanese table, set for a meal, is a delight to the eye: the arrangement and colours of the dishes and the rests for the chopsticks, the presentation of the food itself.

One might well think that the busy Japanese of today would no longer have time to appreciate a beautifully laid table and dishes artistically presented. Not so at all. A Japanese business-man, even if he can take only half and hour for lunch, will go to a little restaurant where he will order a very simple dish and be served with little bowls of tempting food, presented on a wooden tray divided into compartments. At home, in the evening, he will forget the cares of the day and relax, first in the contemplation, then in the consumption, of a meal—perhaps a frugal one—that his wife will most carefully prepared.

The table in accordance with seasons

A Japanese table must express harmony. So the bowls and dishes vary according to the seasons. The Japanese often wonder how Westerners can possibly use the same plates all the year round! They regard it as quite barbaric to watch the autumn leaves falling outside and then turn one's gaze to a bowl decorated in blue! A pale green or brownish bowl, on the other hand, avoids this clash between the colours of nature and the colours of the table. The whole table setting must be carefully designed, moreover, by the mistress of the house, to harmonise perfectly with her flower arrangment and with the rank and character of her guests.

Their profound respect for nature leads the Japanese to respect the seasons too and to eat accordingly. Thus, in February, they will serve strawberries, in spring, young bamboo shoots, in summer, "ayu"—a delicate fish found in certain rivers, particularly at Gifu. The advent of the deep freezer has not changed these habits.

This respect for nature also means that they treat all their nature's produce with infinite delicacy. Japanese cuisine reflects the central preoccupation of all Japanese art—a concern with the essential qualities of things. Both the housewife and the chef take the same pains to bring out the intrinsic qualities in whatever food they happen to be preparing. Presentation is of first importance: a fish, for example, served up in a sauce, bears little resemblance to its natural state. A cook should therefore make an effort to give it back its fishy qualities, by serving it on a bed of sea-weed perhaps, to symbolise its natural element, or by adding a carrot, whose rich golden colour will emphasize the delicate pink flesh of the fish itself. Why should food be a formless mixture? Normally, rice is served on its own; however, there is one particularly attractive dish which consists of a layer of rice on which are arranged—in carefully separated sections—little heaps of ginger, sea-weed, and pieces of roast duck; it looks beautifully fresh and inviting. The guest is free to savour the various components in any order he pleases...

A Japanese cook takes great pains to bring out the individual qualities of foods by skilful use of contrast. Thus a meal will not have just one main dish but rather three or four, of very different consistency and flavour, served together. The guest can take his chopsticks and help himself to alternate mouthfuls, savouring the differences—the salt will bring out the flavour of the sweet. A dish is considered really successful if it brings together food of different consistencies—soft, slippery, firm and elastic.

Japanese cooking, which some Westerners still find barbaric, is thus a matter of great refinement and subtlety; though it must be admitted that it does contain surprises, at first, for a European palate. □

the tea ceremony

■ The "chanoyu", or tea ceremony, is a reflection of the very essence of Japanese art and the Japanese attitude to life. As early as the 8th century, tea was known and used in Japan as a medicine. The "chanoyu" itself is Buddhist in origin. The monks used to drink tea, prepared according to a certain ritual, in order to assist their alertness during meditation. This much esteemed beverage was said to give precision to thought and clarity to reasoning. In the 13th century, the Zen preacher Eisai wrote a pamphlet entitled: "Of the Salutary Influence of Tea", of which the most important part consists of precise instructions for the preparation and serving of the drink.

During the 14th century, the "chanoyu" became a social ceremony, a luxurious court ritual. In their splendid appartments, richly decorated and adorned with precious objects, amid odours of incense, great lords would receive their guests, reclining on couches, and offer them exotic dishes, and tea—of which they were supposed to guess the origin. In the 15th century, Ashikaga Yoshimasa gave up the shogunate in order to enjoy a premature retirement in his villa—the Ginkakuji or Silver Pavilion—in Kyoto, to devote himself exclusively to the art of tea-drinking. At this period, a present of a teaspoon or a tea-bowl was highly valued and considered as a mark of favour. "Daimyo" (lords) and samurai used to give magnificient entertainments and, when their castles were attacked, would die with their tea-bowl—rather than their sword—in their hand. In the 16th century, Toyotomo Hideyoshi gave an elaborate "chanoyu" in Kyoto that lasted for a whole week! Such luxurious orgies provoked an austere reaction. The great master of the art of tea-drinking, Sen no Rikyu, applied himself to devise a set of rules; these rules have governed the tea-ceremony ever since.

The tea-house "Suki", must be set apart, to be reached by a stone-flagged path across a landscaped garden. The tea-room itself must be built of precious woods, decorated with a delicate "kakemono" and a vase containing a few branches of blossom, with tatami on the floor, and, adjoining it, a little kitchen in which to wash the vessels used in the ceremony. The pavilion itself, built of bamboo, with a roof of dried grass supported on slender pillars, should convey an impression of the transcience of things and the fragility of life. The guests should be seated in the tea-room, the most important of them next to the "tokonoma" (alcove). The master of the ceremony should then enter, bringing the utensils with him: a low table of mulberry wood, a lacquer box containing the tea, tea-bowls, a water pot, and a bamboo whisk. After having greeted his guests he should heat the kettle of water—filled with the aid of a long-handled bamboo ladle. The water should then be poured onto the green tea leaves in the tea-bowl and the brew should be stirred with the whisk. The bowl should then be offered to the most important guest, then passed to the others; the company should praise the flavour of the tea, the beauty of the bowl and the arrangement of the "tokonoma". The ceremony over, the master should gather the utensils, take leave of his guests, and leave.

Today many Japanese girls attend special "Chanoyu" schools, for the ceremony is still widely practised. Even large commercial and industrial firms organise "Chanoyu" classes for their women staff. In Tokyo and Kyoto tea ceremonies are held every day, for foreign visitors. But they are often rather perfunctory and the foreigner may well wonder whether this apparently out-of-date ritual really is still part of Japanese life. He will get quite a different impression if he is lucky enough to witness a small private tea ceremony.

The "Chanoyu" is a search for beauty in colour, form, and precise and elegant gesture. The ceremony is a school of self-control, of economy in gesture, of choice of the right object and the right word. The "Chanoyu" brings relaxation to the body and clarity and detachment to the mind. □

town by town

site by site

from abashiri to yokohama

abashiri

"But there's nothing here worth visiting!"—This might well be the immediate reaction of the newly-arrived visitor, not much encouraged by the evident perplexity of his taxi-driver and interpreter when asked about the local sights.

A small Hokkaido town of some 40,000 inhabitants, Abashiri does not impress at first sight; its lifeless streets are lined with very ordinary houses. The harbour is far from picturesque. And yet...

The arrival of winter transforms Abashiri into a fairytale wonderland of ice and snow. The roofs of the houses are transformed by a thick mantle of snow and icicles hang like stalactites from every eve. The harbour opening onto the Sea of Okhotsk is completely iced up. The rock formations round about are transformed into strange sculptures and the lighthouse is stranded amid the frozen waves. It is a strange dreamlike world in which to wander amid the silent snow. The bitter cold may well drive the visitor into a warm and welcoming smoke-filled bar; here his wooden sake goblet will soon be filled by the jovial locals, friendly Japanese on the whole taller than the rest of their fellow-countrymen.

The sight of the Abashiri fishing fleet returning to port, in the summer, will indelibly underline the importance of fish in the national life of Japan. Cranes lift great bales of netted fish from the boats and load them onto waiting lorries—a cascade of colour and a wonderful range of different kinds; the finest and most highly prized are carefully sorted into wooden boxes. They often have exotic names, to match their wondrous shapes...

The gourmet visitor will certainly not leave without feasting on the magnificent local crabs (kani). There are little restaurants, easily recognisable by the huge crab painted up outside, where these lusty crustaceans may be enjoyed. The flesh is firm and delicious; it is the custom to finish with a drink of sake from the empty shell. The locals, contrary to what one might think, never seem to tire of them and may be seen eating crabs at all hours of the day. Abashiri indeed has its charms!

ABASHIRI

Location and access: Island of Hokkaido (n.w. coast), fishing port on the Sea of Okhotsk. Pop. 50,000. Distances by train: Hakodate 660 km, Sapporo 370 km, Tokyo 1,480 km. Airport (19 km): flights (TOA) to Sapporo, Hokadate.
Post code 099-24; telephone code 01524.

Travel agency: JTB Minami Shijo, Nishi Ichi (tel. 4-5205).

Accommodation: Hotel *View Park****, tél. 01524-8-2211, 100 westn. style rooms + 50 trad. style. 15 mins. by taxi from station, fine view across lake. *Abashiri Kanko Hotel* (Ryokan)**, Yobito, tel. 01524-8-2511, 42 rooms, near station.

Festival: Fairytale Ice Festival in mid-February.

AMANO HASHIDATE (p. 92)

Location and access: Island of Honshu (near Miyazu), one of the three classic scenic sights, "san kei", of the Japanese coastline (the others being Matsushima and Miyajima). Tokyo 561 km, Kyoto 122 km, Osaka 159 km.
Telephone code 07722.

Accommodation: Ryokan: *Gemmyoan****, Monju, tel. 2-2171, 32 rooms. One of the most beautiful ryokan in Japan, with fine views of the Celestial Bridge from every room. *Kitanoya Hotel***, Monju, tel. 2-4126, 30 rooms. *Amanohashidate Hotel**, Monju, tel. 2-4111, 31 rooms.

Youth hostel: Amanohashidate, Manai, Nakano, tel. 9-3121, 60 beds.

Souvenirs: What about a nice little dried octopus?

ASO (MOUNT) (p. 92)

Location and access: Island of Kyushu, Kumamoto prefecture. Mount Aso National Park. Beppu 97 km, Kumamoto 50 km. The area can be visited by bus and train from Kumamoto. It is possible to drive there in one's own car, the road is not difficult. To avoid having to go up and down the volcano by the same cable-car it is useful to use a hire-car with driver.

Accommodation: Hotel *Aso Kanko***, Yunotani Choyo mura, Aso gun, tel. 09673/5-0311, 30 rooms and 60 trad. style rooms. Ryokan: *Aso Kogen Hotel****, Aso machi, Aso gun, tel. 09673/5-0211, 18 rooms. *Aso Hakuun sanso***, Aso machi, Aso gun, tel. 09673/5-0111, 114 rooms.

Youth hostel: Aso, Bochu, Aso gun, tel. 4-0804, 60 beds.

Page 86-87 :
*As soon as the fine weather arrives
the parks of Tokyo are full of picnickers,
young and old, who spend the day there talking,
studying, and admiring the flowers.*

akan

■ Akan is the name of a lake, a spa—Akan-Kohan—situated on Lake Akan and dominated by Mount Akan, and of the 87,500 hectare National Park that contains them all. The park may be conveniently visited by taking the Kushiro-Bihoro bus, which obligingly stops at all the beauty spots. Such a bus ride has the added bonus of local colour—or at least the company of many Japanese holiday groups out to enjoy themselves.

The journey from Kushiro to Akan-Kohan is not particularly interesting. It is a good idea to time your arrival at the spa for late afternoon, take a boat trip on the lake and then spend the night at Akan-Kohan. The whole of this densely wooded region is a much favoured holiday resort of the Japanese who flock to it in summer to enjoy its cooler atmosphere. From the first thaw in early spring the lake attracts many visitors who come to marvel at the "marimo", a sort of duckweed which grows in balls up to the size of a large orange and floats prettily on or near the surface of the lake, close to the islands.

Akan-Kohan is also a convenient spot from which to climb Mounts Me-Akan and O-Akan, from which there are magnificent views over the whole region. There are still some Ainu communities (though they are gradually becoming assimilated into the local population) living round the shores of the lake. Lake Akan is the scene of an aquatic festival in October. This is when the Ainu commemorate the legend according to which, many years ago, some of the Ainu people marooned on an island because of the cold lived off food left there during the night by the Koro pokkuru or Kami (spirits) of the forest.

Shortly after leaving Akan in the direction of Bihoro you come to the Soko dai lookout point, from which there is a dramatic view down over the little lakes of Penke and Panke. Some 15 kilometres away lies the crater lake, Mashu ko. It is a very wild spot, difficult of access, and its steeply sloping banks are planted with pines and larches. It is particularly lovely in early spring when the snows are beginning to melt. There is a little islet in the lake; with the snow-capped reflection of Kamui nupuri in the clear icy water it makes an unforgettable picture. It is possible to see down to a depth of 40 metres into the waters of this remarkable lake. Close at hand rises the sulphur mountain, Mount Io san. The sides of this ancient volcano give off powerful sulphurous fumes; it is a curious sight to see the local women cooking eggs in the natural ovens formed by cracks in the ground. The whole region is rich in thermal springs. There is even a hot one, Wakoto onsen, which rises amid the icy waters near the shores of Lake Kutcharo ko. It is pleasant to take a boat trip on this lake in summertime. From the Bihoro toge Pass there is an extensive view—over Lake Kutcharo ko, on one side, and out over the shores of the Sea of Okhotsk, on the other.

AKAN

Location and access: Island of Hokkaido, 75 km north-west of Kushiro and 110 km south of Bihoro.
Post code 0885-04. Telephone code 015-467.

Travel agency: JTB Minmi Shijo, Nishi Ichi (tel. 4-5205).

Accommodation: Ryokan *New Akan Hotel****, Akanko Onsen, tel. 015-467-2121, 140 trad. style rooms. Ryokan *Akan Kanko Hotel***, Akanko Onsen, tel. 015-467-2611, 70 trad. style rooms. Both on shores of Lake Akan.

Festival: Ainu Festival on Lake Akan, 1st Sunday in October.

Souvenirs: "Marimo" (duckweed balls), in glass jars and mounted as key-rings.

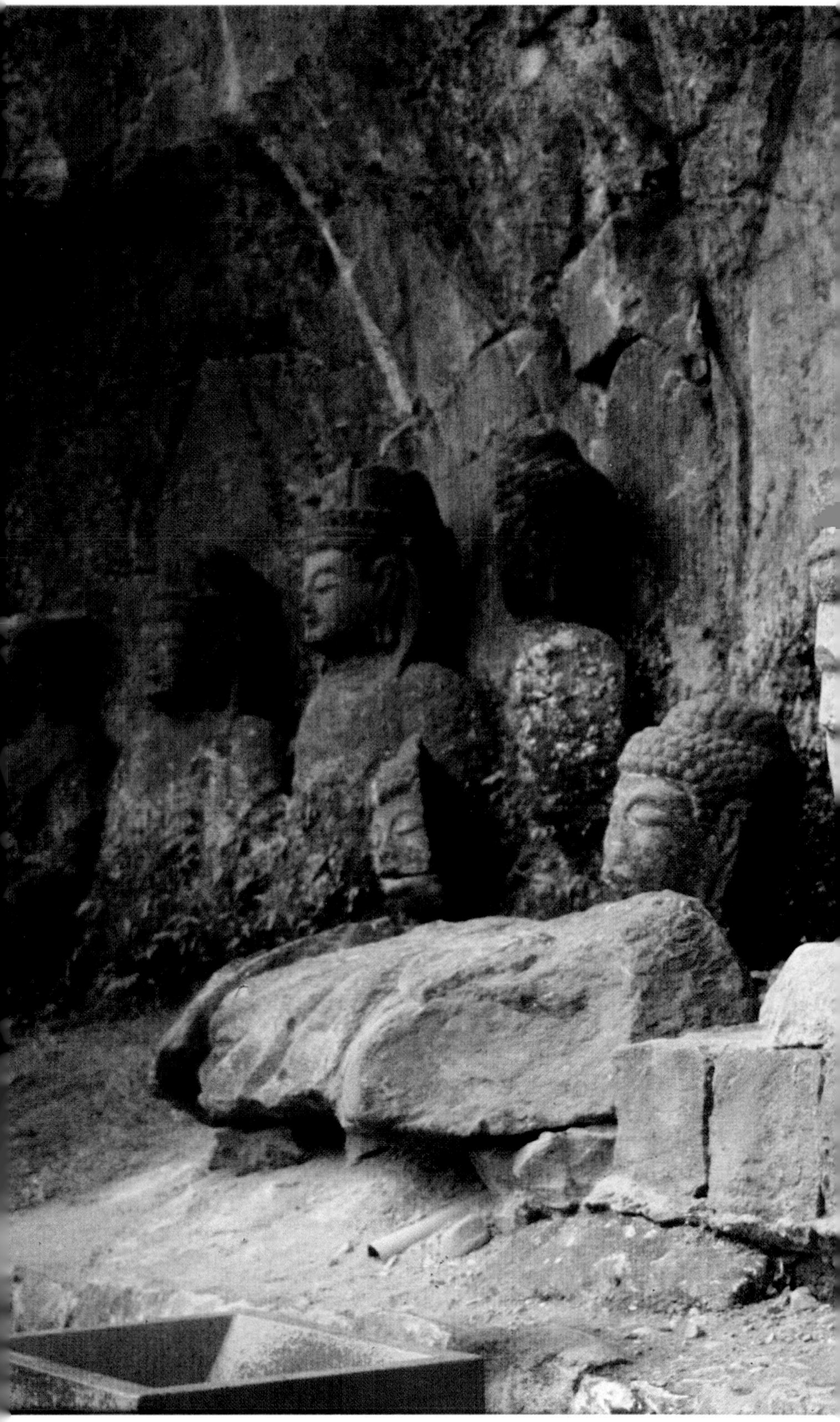

*For ten centuries these great Buddhas
have watched over the peasants
as they work in the ricefields of the Usuki Valley.
Their stone lips bear a distant yet peaceful smile
(photo by C. Salvadé).*

amano hashidate

■ There they stand, looking backwards through their legs, a constant procession of visitors, gazing enraptured at the Bridge of Heaven, Amano Hashidate, from the Kasamatsu Belvedere at Ichinomiya. It is never without its admirers and in summer they swell to a horde, for this is one of the classic "scenic trio", or "sankei" of the Japanese coastline the others are Itsukushima and Matsushima).

Every Japanese hopes to make this pilgrimage to the Celestial Bridge (Amano uki hashi), at least once during his lifetime. This is where the turbulent spirits Izanagi and Izanami (see "*Religion:* Shinto") landed when they descended to earth. The site itself is a very pretty one. A narrow spit of sand three kilometres long, planted with pine trees, separates the Sea of Japan from the Aso Lagoon. At each end lie the two little towns of Ishinomiya and Monju, the latter beyond a bridge which swings to allow vessels to pass through. On each side of the inlet there are hills with charming viewpoints from which to admire the romantic scene, lovely throughout the year in its changing colours. In the mornings, after rain, the pines seem to float amid the swirls of mist which wreath the sandy spit. In winter every detail stands out with the clarity of a fine etching. And there is always the bridge, best seen upside down against the sky, which has inspired poets and painters—notably Sesshu—down the ages.

Starting from Monju, the visitor passes ranks of little souvenir shops (note the curious dried octopuses swinging in the breeze) on the way to the old dark wooden temple, "Chion ji", near the lagoon. There is a cable railway to the belvedere. Ichinomiya can be reached by walking out along the "bridge" or by taking a ferry: there is another cable railway and a chair lift up to the Ichinomiya belvedere.

It is very pleasant to spend a day and a night at Amano Hashidate, except in the height of summer when the whole place is far too crowded for comfort or enjoyment.
(See notes p. 88).

aso (mount)

■ Mount Aso rises from the centre of the Island of Kyushu and is the biggest active volcano in the world —frequently belching out columns of sulphurous smoke. Its last eruption was in 1933. Some 73,000 hectares of the surrounding countryside, including Mounts Aso and Kuji, are classified as a National Park.

Mount Aso is usually visited by bus, starting from Beppu and ending up at Kumamoto. After leaving Beppu there is a steep climb up the slopes of Mount Tsurumi dake, which gives a good view of the town. 60 kilometres of winding road take us through pastures, pine plantations and the small resort of Yufuin Onsen at the foot of Yufu dake, before we reach the huge crater of Mount Aso itself: 128 kilometres in circumference, its sides steep in places, sometimes wooded sometimes grassy. Against a background of ricefields rise the five cones of Mount Aso: Naka dake, Neko dake, Taka dake (the highest, 1,592 metres), Eboshi dake and Kishi dake. This view is particularly lovely in early morning and in the evening, the waters of the ricefields gleam softly and not a breath of wind disturbs the perfect reflections of houses and rice-plants.

After Ichinomiya, a place of pilgrimage to the Aso shrine, the bus climbs again to Sensui where a big cable-car takes us over the lava flow and jagged rocks as far as Narao dake. There a specially fume-proof vehicle takes us to Naka dake, the main crater, from which thick clouds of sulphur rise from time to time. Another cable-car takes us to the little Sanjo jinja shrine where our bus awaits us. We then make our way down the gentle gradients on this side of the volcano, planted with fields of azaleas. One of the highlights of this trip is the view of the pretty little green cone of Takatsuka.
(See notes p. 88)

beppu

■ Situated on a splendid bay, surrounded by wooded mountains, Beppu, an important harbour on the island of Kyushu, is wreathed in steam from its three thousand hot springs. There are large hotels and small ryokan on every slope of the white-capped hills all around. Beppu has the attractions of the seaside, the mountains, interesting excursions and all the amusements of a spa, as well as the curative properties of the waters themselves; no wonder it is a Mecca for holidaymakers as well as sufferers from arthritis and nervous and skin complaints. During the week-end its population rises from 130,000 to 300,000! Beppu is a pleasant, restful spot in which to spend two or three days.

"Hell", or 3,000 springs

The "Jigoku" or "Hell" formed by hosts of hot springs is a remarkable spectacle. Each spring has its own peculiarities—its vigour, colour, smell and temperature (37°-94°C.). The most interesting are some distance apart, so it is a good idea to visit them by taxi. At Tsurumi the waters gush out with a noise like a railway engine, to form a big bubbling pond. Chino ike jigoku is a vast bowl of blood-coloured water, seething at a temperature of 169° Fahrenheit and with a canopy of white steam. The nearby Tatsumaki geyser spurts up regularly every twenty minutes. The loveliest spring is the Umi jigoku, deep seablue, set amidst landscape gardens, palm-shaded and watched over by rows of little stone Buddhas. Alligators splash about in the Oniyama jigoku. There are modest ryokan, with their own springs where the guests take to the waters in steamy privacy in little wooden cabins. According to the statistics however, only some 10% of the 80 million litres of water that gush out every day are actually used. No system has yet been devised to exploit them for domestic heating, though they used by some horticulturists in their greenhouses.

Beppu, mercifully spared by the bombs of the Second World War, has some charming districts hidden away between the harbour and the river—here old wooden houses line the narrow picturesque streets.

After a tour of the Hells and a walk round the town a bathe in one of the "onsen" or a plunge in the hotel swimming pool is an ideal refreshment.

The town is dominated by Mount Takasaki yama (600 metres), a haunt of monkeys. They used to descend in thieving hordes on the market stalls in Beppu. However the present mayor of the town had the brilliant idea of preventing this by setting out food for the monkeys half-way up the mountainside itself. Their antics there at feeding time on Sundays are a popular local attraction!

The Marine Palace Aquarium by the sea shore is well worth a visit if time allows.

In the nearby mountains the largest African Animal Reserve in the Far East has recently been established. The collections of lions, elephants, zebras and so on can be safely and enjoyably visited in one's own car—with a hired F.M. radio to keep in touch with the keepers... just in case.

Beppu is a good starting place for a trip to Usuki, 50 kilometres away in Oita province. Here on a hillside out in the country some sixty Buddhas stand carved out of the living rocksome gigantic, some quite small; some in groups and some alone. It is remarkable to see such a collection of splendid figures in such an out-of-the-way corner of the country. They date from the Fujiwara period, 794-1192 (see: *Japan through the centuries*). The figures are remarkably well preserved, the sculptors' chisel marks are clearly visible in places. Their eyelashes still veil their calm and stony gaze and they retain their original colouring. (Take the train from Beppu to Usuki, then a taxi as far as the Buddhas (8 km).)

There are also day coach-trips from Beppu to Mount Aso. It can also be seen by taking the regular bus from Beppu to Kumamoto (see under Mount Aso).

BEPPU

Location and access: Island of Kyushu, pop. 130,000, Oita Prefecture (pop. 1,200,000). Distances by train: Tokyo, 1,225 km, Oita 12 km, Fukuoka 162 km, Aki Airport (Oita, 39 km), direct flights to Osaka (1 hr) and Nagoya (1 hr).
Post code 874. Telephone code 0977.

Travel agency: JTB Kokusai Kankokaikan Bldg. Kitahama, tel. 22-1272.

Accommodation: Hotels: *Suginoi****, Kankaiji, telex 7734-67, tel. 24-1141, on the hillside, 135 rooms + 471 trad. style. *Hakuusanso****, Kankaiji, telex 7734-62, tel. 23-1151, 14 rooms + 105 trad. ones, on the hillside, views. *Beppu New Grand****, telex 7734, tel. 22-1161, 14 rooms + 97 trad. ones, on the Kajima hillside. *Kamenoi***, 5-17 Chuomachi, telex 7734-75, tel. 22-3301, Town centre. 31 rooms + 57 trad. ones. Reduction for long stay visitors. *Hinago***, 7-24 Akibacho, tel. 22-1111. Town centre, 7 rooms + 65 trad. ones. *Nippaku***, 3-12 Kitahama, tel. 23-2291. Town centre, 7 rooms + 65 trad. ones. Ryokan: *Kodama****, Tsurumi, tel. 66-2211, 56 rooms. *Koraku****, Kitahama, tel. 22-1331, 42 rooms.

Festival: Sakura matsuri, in April.

THE SUGINOI HOTEL

■ *The Suginoi Hotel, at Beppu, is said to be "the most modern, the biggest and the smartest holiday hotel in all Japan". Two gigantic white dominoes rising above a private park, 700 bedrooms, 3,000 beds, enormous lounges, restaurants serving Japanese, European, and Chinese food, theatres, games rooms, pachinko parlours, cinemas, whole floors of souvenir shops, private rooms for 500 to 3,000 guests, hot springs bubbling up in enormous pools surrounded with palm trees, "torii" gateways, statues of Kannon, Goddess of Mercy.*
A whole vast complex, not unduly expensive, somewhat off-putting to a European perhaps, but just the thing for the crowded country that is Japan today.
The Suginoi is a remarkable hotel, an essential part of any visitor's understanding of Japan in the seventies.

*A sacred mountain and a quiescent volcano,
Fuji san has been celebrated by poets
and immortalised by painters down the centuries.*

fuji-hakone-izu (national park)

■ The Fuji-Hakone-Izu National Park covers 1,223 square kilometres and includes the three regions of the same names as well as the seven Izu Islands. Its whole configuration was formed tens of thousands of years ago by volcanic eruptions. No other Japanese national park offers such a wide range of different kinds of natural beauty. It is dominated by the proud cone of Mount Fuji, rising white above its lava slopes, the celebrated Jukai forest, five lakes and charming inland spas, whilst in the distance lie the shores of the Izu peninsula, bathed by the limpid waters of the Pacific Ocean.

The sacred mountain

Mount Fuji, or Fuji san, is the highest (3,776 metres), loveliest and most famous mountain in all Japan. Its volcanic cone has erupted 17 times in recorded history, most recently in 1707 when it covered the city of Tokyo with volcanic ash. It is all-too-possible to visit Japan, not once but many times, without being lucky enough to see Fuji clearly—for the peak is often hidden by mists and cloud. Winter and Spring offer the best chances of getting a good view of it. Ideally it is visible from as far away as Tokyo, Yokohama and from the road that leads from the capital to Kyoto. It is often shown on postcards rising majestically above a diminutive train (shinkansen) that crawls around its base. Its perfectly proportioned outline, its snowy peak are best seen rising in all their splendour from the surrounding plain. The colours of Mount Fuji change throughout the day, to fade away in a pinkish haze in the evening, turning to mauve and then a mineral hue before the shadows of the night envelop it.

Fuji has played a central role throughout Japanese art and literature. It was celebrated as early as the 9th century in the "Manyoshya" anthology. "It can be traced like a golden thread through the fabric of the art and life of the Japanese people." (Fosco Maraini). The engravings of Hokusai (1760-1849), the "Fuji Hyakkei" (the hundred views of Fuji) and those of Hiroshige (1797-1858), "The 36 Views of Fuji", are almost too well-known to need a mention here...

The name "Fuji" is often written in two characters in Japanese, and thus takes on the additional meaning of "unique" and "eternal". In the Ainu language, Huchi or Fuchi is the name of the goddess of the hearth and home. The Japanese never call their mountain "Fuji yama" but always "Fuji san".

Fuji is a sacred mountain. In summer it is climb by hosts of pilgrims; in the old days—during the Meiji period—women were not allowed to go to the very top, they had to stop at the 8th resting place. According to an old Japanese proverb: "He who doesn't climb Fuji once in his lifetime is a madman; he who climbs it twice is a lunatic." Every year 300,000 people climb the sacred mountain. There are six starting places whence six marked tracks lead up to the summit (15-20 km according to the route): Gotemba, Subashiri, Yoshoda, Kawaguchiko, Shoji, Fujinomiya. During July and August there are chalets open along these routes, offering food and lodging. It takes from 6 to 9 hours to reach the summit (4 to 5 hours to return) climbing first through a belt of lush vegetation, smelling sweetly of pines and scented bushes, finally crossing a stretch of ash and lava near the top. Most pilgrims make the ascent (leaning heavily on their sticks, which they dutifully have engraved with an appropriate emblem at each stopping place) during the afternoon, spending the night in a chalet, in order to watch the sunrise from the summit. The climb is easier nowadays, thanks to roads which bring coachloads of climbers as far as the 5th and 6th stopping places. There are six minor peaks set around the edge of the crater, which is 500 metres across and 80 metres deep. Its inside slopes are steeply shelving blackish cliffs.

There is an excellent road around the lower slopes of Mount Fuji which makes possible a most interesting hundred kilometre drive, taking in the five lakes in the region. Leaving from Gotemba, Lake Yamanakako, a small lake and the highest of the five, is soon reached. The Japanese delight in boating here in summer and skating in winter. In winter too there is the picturesque sight of people fishing through holes in the ice. The second fortnight of April is the cherry-blossom season on the banks of Lake Yamanakako. On the 1st of August there is a special festival when the calm waters of the lake are alight thousands of floating lanterns. There are fine views of Mount Fuji itself from the road around the lake. Fuji Yoshida and Kawagushiko (a resort on the banks of Lake Kawagushi) are tourist villages and busy centres for excursions into the lake district, as well as being the most popular starting points for the climb up Mount Fuji. On the east bank of Lake Kawagushi rises Mount Tenjo san. There is a cable-car to the summit, from which there is a splendid view of Fuji reflected in the lake.

There is a toll-road from Kawagushiko to Komitake—the highest point on Mount Fuji accessible by car (2,300 metres). From there it is another 4 hours' walk to the summit. There are look-out points along the track which give marvellous views out over the Alps.

The striking temple of the "Soka Gakkai"

Shortly after Narusawa, famous for its bat-haunted caves, there is a look-out at Koyodai (hill of the maples) from which Lakes Shaiko and Shoji can be seen, as well as the celebrated "Jukai" forest. Lake Motosu, which never freezes over, is beloved of the Japanese for its beautiful blue waters and its excellent trout. Close by is the impressive Shiraito no taki waterfall, where "threads of silver" cascade down in a fall 130 metres wide. Just before Fujinomiya there is a striking view of the Taideki ji temple, the present-day headquarters of the "Soka Gakkai" sect (see "*Religion*"). The first temple built here, at the end of the 12th and beginning of the 13th century, was the seat of the sect of Nochiren sho: "This temple seems to re-create in concrete the volumes and rhythm previously determined by its former wooden structure." (Elisseeff). The temple is open to visitors; it is remarkably richly decorated. At Fujinomiya the goddess of the volcano, "Konohara Sakuya bime" is venerated in the little Sengen temple.

Hakone

Nestling amid the mountains the whole verdant Hakone region that lies between Mount Fuji and the ocean has nowadays become one of Japan's most important holiday centres and a place where thousands go to enjoy the thermal springs for cures or just for pleasure. In the old days Hakone used to be a staging post on the road to Todai do. Every one passed this way—"daimyo" (lords) in splendid costumes, on horseback or carried in chairs by porters, escorted by "samurai"; archers and retainers carrying banners; religious processions, pilgrims, "ronin" (masterless samurai) in search of adventure, poets and minstrels. The Tokugawa shoguns had the idea of erecting a barrier here. The shoguns obliged their vassals to spend a part of every year in Edo (Tokyo); when they were not actually resident they had to leave their wives and families behind there. The Tokugawa shoguns thought—with perfect logic —that if their vassals were planning any sort of revolt they would try to get their families out of the capital, rather than let them become hostages. So every party passing through Hakone was most carefully searched...

The Tokai-do has an importance place in Japanese art and literature. The Noh plays (see "*Theatre*") often have this road as their setting. The 19th-century writer Ikku, had the hero of his novel "Hizakurige" travel along this road. Both Hokusai and Hiroshige delighted in depicting its 53 different "stages".

Moto-Hakone is a pretty resort on the shores of Lake Ashi. It has numerous hotels and ryokan. Throughout the year there are boats plying on the lake between the various resorts. By taking one of them it is possible to see the little Hakone Jinja Temple, built in 757 and later used to confine Minamoto Yoritomo, defeated in 1180 after his struggle against the Taira (see under "History of Kyoto"). The temple has a lovely "torii" (gateway) standing in the waters of the lake itself. There is a cable-car at Hakone-en, up to the summit of Komaga take—from where there is a good view on a clear day, across to Mount Fuji and down over the Izu Peninsula. At Nino Daira, on the Tokai do road not far from Hakone, just before Kowakidani, the Open Air Museum of Hakone was opened in 1969. Here sculptures are displayed in a vast garden designed by Inoue Bukichi, himself a sculptor. As well as works by Japanese artists there are pieces by Archipenko, Bourdelle, Giacometti, Moore and Zadkine. "A perfect harmony of sculpture and natural environment." (Elisseeff). Most of the villages in this area have their own thermal springs: Yumoto onsen and Tonosawa are good for nervous diseases and rheumatism. Miyanoshita is generally considered the leading spa. Its Hotel Fujiya, built in 1878 in Western style, is set in a magnificent Japanese garden with a lake, a waterfall and a tea house; it is well worth a visit.

The Izu Peninsula

The mountains of Hakone project out between Suruga Bay, to the west, and Sagami Bay, to the east, to form the Izu Peninsula. Its mild climate, the wooded mountains sloping steeply to the sea, its hot springs and its beaches all combine to make it an extremely attractive holiday area for the Japanese, particularly during the winter months. A good largely scenic road runs right round Izu, giving fine views of the sea and the island of Oshima, on the one side, and of Mount Fuji, on the other. Atami, the principal resort, is proud of its fine museum (paintings, sculpture, prints and "sutras") and of having the biggest camphor tree in Japan (at the Kinomiya shrine). The road from Atami to Shimoda passes through numerous little spas and watering places. Thanks to its hot springs Atagawa possesses beautiful tropical gardens. Under the Treaty of Kanagawa, Shimoda (together with Hakodate) was one of the Japanese ports opened to American traders in 1854. The Gyokusen ji temple at Shimoda was the residence of Townsend Harris, the first American consul in Japan, it contains relics of the period. The little harbour and the beaches amid the rocks are quite charming. From Cape Iro, to the south, the 7 islands of Izu can be seen. Boats ply across from Shimoda to these little coastal islands, mainly to Oshima, of volcanic origin and noted for its extremely mild climate and its forests of camelias, in flower from November until April. In the centre of the island rises Mount Mihara, an active volcano which permanently emits thick columns of smoke.

FUJI-HAKONE-IZU (NATIONAL PARK)

Location and access: A park of 122,000 hectares, one of the most visited in all Japan. Very easily reached from Tokyo. Enquire at JTB offices. If travelling independently take the "shinkansen" and get out at Odawara (84 km, 45 min.), or at Atami (95 km); buses run from both towns to Hakone. Oshima Island is accessible by air from Tokyo.
Post code 401-03. Telephone code 05557.

Accommodation: Kawaguchiko: *Fuji View****, 511 Katsuyama Minami Tsuru gun, tel. 055583-2511. 68 rooms + 8 trad. style. Hakone: *Fujiya Hotel****, 359 Miyanoshita, Hakone-machi, tel. 0460/2-2211. 20 min. by car from Odawara railway station. Splendid Japanese gardens. Traditional-style building. 190 rooms. *Hakone Kanko Hotel****, 1245 Sengokuhare, Hakone, tel. 4-8501. 110 rooms. Very lovely view of Mount Fuji. 40 min. by car from Odawara. *Yumoto Fujita Hotel****, 256 Yumoto, Hakone, tel. 5-6111. 15 min. by car from Odawara, very lovely view, 100 rooms. + another ten hotels of all categories. Atami: *New Fujiya*****, 1-16 Ginza cho, tel. 81-0111. 318 rooms (139 trad. style). *Atami Fujiya****, 13-8 Ginza cho, tel. 81-7111, 171 rooms (129 trad. style).

Festival: 31st July–2nd August, the Hakone Shrine Festival at Moto-Hakone. Fireworks. A fairyland of thousands of lanterns drifting on the lake.

fukuoka

■ A commercial and academic centre, situated on the northern coast of Kyushu, Fukuoka will not retain the visitor for more than a day. This will be time enough to visit the famous Dazaifu shrine and then to wander down to the bay where he can gaze out to see and imagine the Mongol fleet arriving here...

The site of present-day Fukuoka (formerly known as Najima) was certainly the scene of some of the earliest contacts between Japan and China: these go back to the very beginning of the Christian era. During the 9th century Chinese civilisation entered Japan through Hakata which city maintained active links with the mainland. It was in the Bay of Hakata, on the site of modern Fukuoka, that a Mongol fleet, struggling in in the teeth of a storm, vainly attempted a landing in the year 1274. When the Mongols returned in 1281 (see: *Japan through the centuries*) they found themselves faced by hastily-erected coastal fortifications; their fleet was routed by a "kamikaze" (wind of the gods). During the 17th century Najima built itself a castle, in ruins today, and took the new name of Fukuoka. In 1889 Fukuoka joined up with the neighbouring town and flourishing commercial centre, Hakata, on the other side of the river Nakagawa.

Dazaifu Temman gu, or Sugawara jinja, one of the most famous shrines in Japan, lies some 16 kilometres from Fukuoka, (change at Futsukaichi if travelling by train). It was founded in memory of Sugawara Michizane, deified under the name of Kanko, a patron of the arts and of calligraphy. Michizane was an important figure in Kyoto during the 9th century. Intrigues fomented by members of the Fujiwara clan caused him to be exiled to the province of Dazaifu, where he died in 903. The shrine, built in the 10th century, was reconstructed during the 16th and restored in 1950.

It stands in an unspoiled natural setting and its beauty is heightened by the contrast of its brilliant red buildings against the surrounding vegetation. In the spring (February and March) its gardens are full of plum blossom; in June crowds come to admire the "iris pool". In front of the main building stands an enormous camphor tree and the "flying plum tree" whose seed is said to have been blown here by the wind all the way from Michizane's garden in Kyoto.

At Kashii, 8 kilometres north of Fukuoka, there is a little shrine, dating from 1800, which commemorates the spot where the Empress Jingu is said to have taken ship for Korea. There is a legend attached to this event which is said to have taken place in the year 800, according to the Japanese calendar, or in 350 according to the Chinese. The Emperor Chuai having died suddenly, his widowed Empress herself directed the expedition against Korea. Being pregnant, and feeling that her time was approaching, she placed a large stone in her belt to prevent her baby being born; she was thus able to lead the expedition. After a very stormy return crossing, she landed at Kashii, removed the stone and gave birth to the Emperor Ojin. The whole of the Genkai coastal park, which extends east and west of Fukuoka, is served by buses (enquire at the JTB office). The coastal road affords many pretty views of the rocky shoreline, beaches and offshore islands.

Location and access: Island of Kyushu. Pop. 1 million. Terminus of the "shinkansen" service (Hakata station), Tokyo 1,176 km, (7 hrs), Osaka 624 km, (4 hrs).
Airport at Itazuke, 7 km.
Post code 815. Telephone code 092.
Travel agency: JTB, 1 Tenjin chuo ku, tel. 771-5931.

Accommodation. Nishitetsu Grand Hotel****, 6-60 Daimyo -chome, Chuo-ku, tel. 771-7171, 308 rooms, central, 10 min. from station by taxi. *Hakata Miyako Hotel****, 2-1-1, Hakaaeki Higashi, Hakata-ku, tel. 441-3111, 270 rooms, opposite the "shinkansen" station. *Hakata Tokyu Hotel****, 1-16-1 Tenjin, Chuo-ku, tel. 781-7111, 270 rooms, 5 min. from station and 15 min. from airport. Hotel Station Plaza**, 2-1 Hakataeki-mae, Hakata-ku, tel. 413-1211, 250 rooms. *Hotel Takakura**, 2-7-21 Watanabe-dori, Chuo-ku, tel. 731-1661, 60 rooms, quiet residential area. Ryokan: *Daichokaku****, 1-21-57, Heiwa, Minami ku, tel. 531-2165. *Hakata Kanko Hotel Taiseiso****, 5-4-11 Nakasu, Hakata-ku, tel. 281-0837, 50 rooms, luxury class.

Festivals: 3-4 May, Hakata Dontaku, procession through the streets. 1-15 July, Hakata Yamagasa, at Kushida Jinja.

gifu

"My friend gave me three umbrellas
The first as shelter from the sun
The second as shelter from the rain
And the third to wave to him".

The city of Gifu was completely rebuilt after the earthquake of 1891, and again after the second World War. It is an important commercial and industrial centre, quite close to Nagoya. It rather lacks charm today, but it is still famous for its umbrellas, its lanterns and above all for fishing with cormorants.

All the pretty sunshades and umbrellas in Japan—exported all over the world—are made in Gifu. It is delightful experience to visit an umbrella workshop, often in a modest wooden house, for it is a largely domestic industry still. The whole family—parents, children, uncles and aunts—will be found bent over the skeletons of umbrellas in the making, sticking on the paper and decorating the finished articles. A venerable old man sits cross-legged under a great canopy of orange umbrellas painting another magnificent one in scarlet and black. For the Japanese an umbrella is a luxury but one in which they delight to indulge!

Those wonderful red lamps hanging outside the restaurants, those hundreds of little lanterns that swing from the eaves of the temples, they were all made here in Gifu. Lanterns have been used in Japan for at least a thousand years. But as recently as a century ago the lanterns of Gifu were little known. It took a full day to transport them, by cart and boat, to the important market in Kyoto. The situation was transformed by the opening of the railway, in 1889, and now Gifu has a manufacturing monopoly.

These folding lanterns are used a great deal in festivals, as well as to decorate temples and restaurants. There are spherical ones, oval ones and cylindrical ones; some are made to hang up while others stand on three legs; some are plain, some are

CORMORANT FISHING AT GIFU

■ *The fishing takes place at night, by torchlight... The man in charge, easily distinguished by his hat, has twelve cormorants... each with a ring round the base of its neck. These rings, made of bone, metal or whalebone, are big enough to allow the bird to swallow the little fish from which it lives, but small enough to prevent the bigger ones from passing down into its stomach. The birds are encased in a sort of wood and whalebone contraption, with a cord attached, which allows them to dive but not to escape. The head fisherman throws his twelve cormorants into the water, one after the other, keeping careful hold of their "reins" in his left hand. The cormorants dive for the fish, attracted by the light. The fisherman pulls the birds back to him, opens their beaks with his left hand, and extracts the fish with his right... Each bird brings back on average four to eight big fish each time, thus up to a hundred and fifty may be collected in an hour... These cormorants are tended most lovingly, a traveller tells me that he has seen them protected from insect and mosquito bites by nets during the summer. Each bird has a number according to its age and seniority. The n° 1 bird enjoys a place of honour on board and wears a correspondingly lordly expression...*
(From a letter sent by General Palmer to the "Japan Times", 17 July 1889.)

* *Cormorant fishing has changed somewhat over the years, but it remains the great attraction of Gifu during the summer months.*

decorated; their sole purpose is to be objects of beauty and delight. In 1951 the sculptor Isamu Noguchi invented what he called "luminous sculptures"—lanterns without wooden rings top and bottom and made of thick white paper rather than the traditional tissue. But these white lanterns conjured up associations of spirits returned from the dead and were a complete flop in Japan. (Though they are to be seen decorating every house during the "Bon" —Festival of the Dead—when they summon the spirits of the departed.) It is fascinating to visit the Ozeki Lantern Factory and watch their craftsmen dextrously sticking the paper to the bamboo framework and applying the decoration with a tracing or a fine brush.

An evening on the river

But the great attraction of Gifu for the visitor is the cormorant fishing. Until a few years ago it was a "genuine" trade; it has now become more or less entirely a tourist entertainment, extremely popular with the Japanese themselves. Reluctant to take long holidays, the Japanese love weekending and Gifu is thronged with visitors during the fishing season—from May to September.

At about five o'clock in the afternoon the buses start unloading their eager sightseers near the banks of the river Nagara. Dressed in their "yukatas" (cotton kimonos) they press eagerly towards the waiting boats where they happily squat down in front of little tables to dispose of a copious meal served on a tray. The boats cast and make their way a short distance upstream where they cast anchor. For the next three hours there will be eating and drinking —plenty of wine and sake are on hand—and the atmosphere will become distinctly cheerful!

At about eight o'clock the cormorant fishing boats appear. They slowly drift past the anchored vessels, with all their paraphernalia of lights and tethered cormorants, they make their way slowly back... and that is the end of the show. It is all a very pretty sight, though perhaps slightly disappointing to anyone who had expected the traditional "genuine" thing.

GIFU

Location and access: Island of Honshu, pop. 400,000, capital of Gifu-ken Prefecture (1,850,000). Distances by train: Tokyo, 380 km; Nagoya, 37 km; Kanazawa, 226 km. Post code 500. Telephone code 0582.

Travel agency: JTB, Chamber of Commerce Building, Kanda Cho, tel. 64-1991.

Accommodation: Gifu Grand Hotel***, 648 Nagara, tel. 33-1111. 82 western-style rooms + 75 trad. style. *Nagaragawa****, 51 Ukaja-Nagara, tel. 65-4321. 50 western style rooms + 47 trad. style. (Both hotels are on the Nagara river, 15 minutes by taxi from Gifu station.)
Ryokans: *New Nagarakan****, Nagaragawa Onsen, tel. 65-4321, 49 rooms. *Juhachiro***, Minato machi, tel. 65-1551, 50 rooms.

Restaurants: Banshokan** (Chinese and Western cuisine) 2 Omiya cho, tel. 62-0039. *Gifu Kaikan*** (Japanese and Western cuisine) 39-1 Tsukasa cho, tel. 64-2151.

Youth Hostels: Gifu, Kami Kanoyama, tel. 63-6631. Kodama so, Ken ei, Nagara Fukumitsu, tel. 32-1922, 150 beds.

Festivals: Naked Pilgrims' Festival, 10th December.

Souvenirs: lanterns and umbrellas.

HAKODATE (p. 102)

Location and access: Third largest city of Hokkaido, the island's main port. Airport 7 km, direct flights to Tokyo (50 min.). Tokyo 830 km, Sapporo 265 km, by train. Post code 040. Telephone code 0138.
Travel agency: JTB, 20-1, Wakamatsu cho, tel. 22-4185.

Accommodation: Hakodate Kokusai Hotel***, 5-10, Ote-machi, tel. 0138-23-8751, telex 9926-04. 100 rooms (inc. 15 trad. style). 5 min. walk from the station; 20 min. by taxi from the airport. *Hotel Hakodate Royal***, 16-9, Omori-cho, tel. 0138-22-9181, telex 992-735. 70 min. 5 mins. by taxi from station.

Festival: Snow and Ice Festival at the end of January.

hakodate

■ The visitor taking the ferry to Hokkaido, across the Tsugaru Strait, usually arrives at Hakodate in the late afternoon. This is a good time to get a favourable first impression of Hokkaido, the wild and distant island of which he has heard so much. Hakodate, rather nondescript by day, in the evening becomes a picturesque fishing port. What seemed rather dull wooden houses now come to life as friendly bars and cafes, where the visitor is soon made warmly welcome by the local sailors and fishermen who will ply him with quantities of sake and tell him all about their tough lives on and around the island. Next day it's a good idea to take the cable-car (easily reached by taxi) up to the lookout point at Hakodate yama, an extinct volcano that rises up steeply behind the town. The view—over the isthmus of Hakodate, the Tsugaru Strait, northern Honshu and the Sea of Japan—is very beautiful. The regional museum in the town contains interesting local archaeological collections as well as relics of the "Ainu" civilisation.

The European-style Goryokaku fortress is a reminder that Hakodate was one of the first Japanese towns to be opened to foreigners after the Treaty of Kanagawa (1854). It was in this castle, in 1869, that Enomoto Takeaki, a supporter of the shogun, took refuge and kept the Imperial armies at bay for six months. The Commission for the Colonisation of the Island of Hokkaido was established here in Hakodate in 1869.

A few kilometres from the city there is a Trappist convent, founded in the last century by a group of French nuns; it is famous for the butter, cheese and biscuits that are made there—they tend to be scarce commodities in Japan. Six kilometres from the convent there is a hot spring, Yunokawa onsen; its waters, that gush from the ground at a temperature of 30°-60°C., are well-known for their curative properties...
(See notes p. 101)

himeji

■ There is one great sight at Himeji: the castle, "Himeji-jo". Perched up on a hill, dominating its surroundings, this "Castle of the White Heron", seems ready to take off into the sky. It has a lofty, graceful five storey donjon, roofed with grey tiles, and eaves upturned at the corners, with dolphins on them.

After various battles it fell into the hands, successively, of the Akamatsu, Kodera, Toyotomi Hideyoshi and Ikeda Terumasa families, who all contributed to its building. It was completed in 1617. It is generally considered the most beautiful castle in Japan and is a good example of a "spiral" construction. There are no drawbridges over the moats, simply gangways which could be destroyed if the castle was attacked. Within the moats there are successive layers of fortifications, plentifully provided with loopholes, and communicating with each other by a series of massive doors not one of which is aligned with another. In the very heart of the castle, surrounded by courtyards, guardrooms and storehouses, stands the palace where the lord and his family lived, in the shadow of the fortified donjon—their last refuge. "Himeji Castle shows how elegantly these fortresses could be transformed into magnificent dwellings, places whose beauty ranges from the transient splendours of plum and cherry-blossom to the fierce proud tigers painted on the golden panels of the 'fusama' (paper screens)." (Elisseeff.)

The donjon is now a museum with a fine collection of painted screens and armour—splendid helmets complete with horns and moustaches. It also affords a magnificent view of the countryside around.

Location and access: Island of Honshu. Pop. 450,000. Distances by train: Tokyo 600 km, Osaka 86 km, Kyoto 125 km.
Post code 670. Telephone code 0792.

Travel agency: JTB, Sanwa Bldg., Arami machi, tel. 22-2141.

Accommodation: Hotel: *Himeji New Osaka***, 198-1 Ekimae cho, tel. 23-1111, 38 rooms (+ 4 trad. style rooms). Ryokan: *Banryu*, Shimodera machi, tel. 22-5655, 17 rooms.

Festival: Castle Festival, 22 June.

"Go in peace, so that this crime may never be repeated".
Behind the Cenotaph stands the shell of the Industrial Exhibition Hall, a poignant reminder of 6th August 1945 in Hiroshima (photo by C. Salvadé).

hiroshima

■ On 6 August 1945, Hiroshima was waking up to a perfect summer's day... then at 8.15 came a blinding flash, a deafening roar and unbearable heat. The first atomic bomb had fallen from a silent sky to destroy this city of 400,000 and to leave 78,150 victims in its wake. Thousands more were badly injured, others suffered the effects of radiation from which they died often years later. Those who survived had had an unforgettable glimpse of hell on earth...

Today, more than 30 years later, the reconstructed city of Hiroshima still remembers. Every year there is a great commemorative ceremony. According to Buddhist belief the spirits of the dead return at this time to spend a few days in their former homes. Thousands of coloured lanterns, brightly burning symbols of the departed souls, are floated down the river, watched by silent crowds gathered in the Peace Memorial Park.

At one end of the park is the shell of the Industrial Exhibition Hall, the only building to survive the holocaust, now known as the Atom Dome. In a line with it stand the Cenotaph and the monument of the Eternal Flame, designed by Tange Kenzo, the distinguished Japanese architect who directed the rebuilding of the city. The Fountain of Prayer and the Peace Memorial Museum (open 9.00-4.30) lie beyond.

A visit to the Museum (also the work of Tange Kenzo) is essential, though painful, in order to understand the ghastly tragedy that occurred here. The display is complemented by a short film. The visitor will no doubt wish to pay his respects at the Cenotaph, which bears the names of the bomb victims and the words: "Remain at peace so that this mistake may never be repeated."

The park is well laid out and planted with lovely trees; it is a calm a peaceful place, popular with the people of Hiroshima. There are se-

A LITTLE RESTAURANT IN HAKODATE

■ *A wooden house just like all the rest... A sliding door that lets out gusts of warm air, enticing smells, snatches of music. Inside, behind a long, semi-circular counter laden with all sorts of fish piled up on heaps of ice, leaning up against a dresser filled with little blue bowls and flasks of sake, stands the chef, perched on high "geta" and dressed in a short kimono, with a wide printed ribbon round his head. The customers sit up at the counter on high stools. A waitress quickly brings a wooden beaker with salt round its rim and fills it generously with sake. Orders are given, fish are skilfully impaled on spits and set to grill... a few minutes later succulent scallops will be served from little long-handled shovels, or dragon-mackerel, and the customers will soon be plying their chopsticks to separate out the bones... More sake is called for, and more beer, it gets steadily warmer, the plangent notes of the "samisen" are lost amid the laughter. A sailor says goodnight, and leaves, swaying gently...*
(Such little restaurants are common in Hakodate and all over Hokkaido. Unfortunately their names are written up in Japanese characters only, so it's difficult to recommend individual ones; they are all pleasant and welcoming to visitors.)

veral fine sculptures such as the Mother and Child in the Storm, as well as other commemorative stones. It is a happy place, full of laughing children and young mothers and babies, but the sight of older people, their faces lined with suffering, lost in recollection near the Cenotaph, serves to remind us of what we would perhaps rather forget...

Yet there are other aspects to Hiroshima, which was founded in 1594 by Mori Terumoto, who built the castle of Ri jô (the carp) in the middle of the river, on the isle of Hiroshima. The castle was destroyed by the atomic bomb but was rebuilt in 1958; it houses in its five-storey donjon a Museum of Hiroshima and its region, with displays of ancient pottery, maps, manuscripts and weapons.

From the top of the castle keep there is a view over the city and its port, a modern city of nearly a million inhabitants, capital of a Prefecture and situated on the industrial plain by the Inland Sea. Naval shipyards and factories producing cars, chemicals, textiles and food products illustrate the dynamism of people who have not forgotten the past but live resolutely in the present and look towards the future.

It is worth spending at least a day in Hiroshima. The gardens of Shukkei en, a classic landscape designed and created in 1620 for Asano Nagaakira, lord of Ri jô, are delightful. Visitors will be enchanted by its little bridges, pools, and islands—a miniature copy of Lake Si Huon, in China near Hang Chow—not to mention the charming tea houses...

HIROSHIMA

Location and access: South Honshu, on the Inland Sea. Tokyo 873 km, Okayama 165 km, Fukuoka 180 km, Pop. 850,000. Capital of Hiroshima Prefecture (pop. 2,700,000). Rail links to Tokyo (5 hrs 10 by shinkansen), Fukuoka (1 hr 45), Okayama (1 hr). Airport 7 km: air link to Tokyo in 2 hrs 10. Many shipping services.
Post code 730/733. Telephone code 0822.

Information: Departmental Tourist Authority, 10-52 Moto machi, tel. 28-2111.

Accommodation: Hotels: *Hiroshima Grand*****, 4-4 Kami Hachobori, tel. 27-1313, 400 rooms, in the city centre, 5 min. by car from the station, 20 min. from the airport. *Hiroshima Kokusai****, 3-13 Tate machi, tel. 48-2323, 85 rooms, splendid view of the city from the revolving restaurant on the top floor. *Hiroshima Station****, in the JTB station building, tel. 62-3201, telex 652-993, 150 rooms. *Hiroshima River Side****, 7-14 Kaminobori cho, tel. 281251, overlooking the river, near the station. *New Hiroden***, 14-9 Osugo cho, tel. 63-3456, 368 rooms (460 beds), business hotel, near station and city centre.
Ryokan: *Itaya***, Hashimoto cho, tel. 21-2391, 16 rooms. *Kakusuien***, Fukura machi, tel. 48-1221, 17 rooms. *Mitakiso**, Mitaki cho, tel. 37-1402, 21 rooms.

Restaurants: Western cuisine in all the big hotels. *Lira*** (West. cuis.), 1-4-7 Kamiya cho, tel. 47-2151. *Suehiro Hiroshima ten*** (West. and Jap. cuis.), 1-21 Tate machi, tel. 47-7175. *Teien Restaurant Hanbei*** (West. and Jap. cuis.), 8-12 Honura cho, tel. 82-7121.

Festivals: 15 July, "Kangen sai" at the Itsukushima Shrine at Miyajima (22 km). 6 August, Peace Festival in the Peace memorial Park.

hokkaido (island of)

■ "A journey to Hokkaido is fraught with difficulty. It is not advisable other than between mid-May and mid-July. Winters in Hokkaido are bitterly cold, summers are extremely hot and rendered most unpleasant by hordes of mosquitoes. There are few good inns. Rickshaws, "kuruma", are to be found in only a few places. The horse is the usual means of transport; it is advisable to take one's own saddle..." Felicien Challeye, 1915.

Today Hokkaido is only an hour's flight from Tokyo; the railway tunnel under the Straits of Tsugaru is due for completion in 1980. All the large towns on Hokkaido have daily flights to the capital, Sapporo. The remotest places of interest to the visitor are served by train and bus. The island has become popular with skiers in winter and with those seeking a pleasantly cool climate in summer. The shores of its mountain lakes become favoured honeymoon resorts; the hot springs, "onsen", are magnets for visitors wishing to combine cures with agreable excursions to the Akan, Daisetsuzan and Shikotsu Toya National Parks.

Hokkaido was for long the poor relation of Japan. It was where undesirables were sent; it was the refuge of defeated clans like the Fujiwara. It was in fact the Japanese Siberia. The first poor families who voluntarily went to Hokkaido to seek a better living usually left their native Honshu with heavy hearts. Hokkaido was the end of the world, a barbarous place with no traditions, no temples, an abandoned island apart from its native Ainou.

The origins of the Ainou people are lost in obscurity. They used to live in northern Honshu, whence they were gradually expelled and driven across to Hokkaido.

Physically they are tall, fair-skinned, hairy and have only slightly slanting eyes. The Japanese used to call them "dog men" or "bear men". The older men still often wear beards. The Ainu women used to be famed for their moustaches. However, their ideas about beauty have changed and their upper lips are now

*Perhaps the Japanese still offer prayers
to the Sun Goddess Amateratsu,
remote ancestor of the emperors of Japan...*

carefully plucked. The Ainu still speak their own language, quite different from Japanese. They used to be a people of fishermen and farmers who kept themselves apart. Today they are merging into the general population of Hokkaido. Some still live in native areas, reserves such as Shiraoi, where they maintain their old traditions and crafts—and are a major tourist attraction. Yet a "Hokkaido Liberation Movement" is gathering momentum at the moment; it is said to have Ainu backing.

Until the Meiji Restoration in 1868, Hokkaido remained neglected. The governments only concern there was to keep the Ainu firmly under control. Fear of Russian invasion led to the building of fortifications, around 1812. Similar fears were responsible for schemes of deliberate colonisation, after 1868. A Commission was set up, first at Hakodate, then at Sapporo in 1871, with responsibility for administration and settling Japanese immigrants on the island. Foreign technicians were called in, especially Americans.

A foreigner landing in Hokkaido will scarcely believe he is still in Japan. The landscape is quite different. Here there are no closed in horizons, no wooden houses with blue tiled roofs reflected in the flooded ricefields—still so characteristic of the non-mountainous agricultural areas of Honshu, Shikoku and Kyushu. Hokkaido is a land of wide horizons, high mountains covered with huge forests (75 % of the land surface), big lakes, a rugged landscape ravaged by cold and snow for eight months of the year. The temperature sometimes drops here to minus forty degrees Centigrade and every winter two or three peasants are eaten by bears! The houses on Hokkaido are sturdy rather than beautiful, their thick walls carry metal roofs painted in garish colours. They often stand out in the fields beside the forage silos.

Yet this large island (second largest in Japan), with an area of 78,000 square kilometres, lying between the Pacific Ocean, the Sea of Japan and the Sea of Okhotsk, is beautiful in winter under its blanket of snow and in spring when the thaw sets in on the lakes. It is still only sparsely populated (5,330,000 or 66 per square kilometre) but its people are young and active. Apart from rice, crops consist mainly of rye, sugar beet and forage; only one harvest a year is possible. Grazing is being developed along with forestry. Fishing occupies the coastal population and yields 60,000 tons a year. The future seems assured by the islands mineral wealth in iron, gold, chrome, oil and natural gas. The cities are all expanding rapidly. The Winter Olympics of 1972 made Sapporo world famous.

The world may just know of Hokkaido now, but with the Japanese it has become a mania. They flock over to visit an island that is so different from the rest of the country. It now has modern hotels as well as plenty of "ryokan"—inns of the traditional Japanese type. Western visitors are still comparatively rare, though some do come over during the summer. But winter is really the ideal season for a visit to Hokkaido; even a short stay then can be very enjoyable indeed. Warm clothes are essential of course, but the island really seems to be most itself at his time of year —whilst the pleasure of bathing in an "onsen" (hot spring), when the ground all around is white frost, is something very special!

HOKKAIDO (ISLAND OF)

Name: Hokkaido used to be known as "Yezo" or "Ezo" which means "the way to the north sea".

Location and access: Hokkaido is the northernmost of the large islands of the Japanese archipelago (pop. 5,338,000—i.e. 66 per square kilometre—the Japanese average is 303). The airport, Chitose, is 15 kilometres from the capital, Sapporo (pop. 1,200,000); the flight from Tokyo takes an hour, from Osaka an hour and threequarters. When the rail tunnel is completed between Honshu and Hokkaido (40 kilometres, the longest in the world) Sapporo will be a mere 6 ½ hours by train from Tokyo.

For information about tours and trips in Hokkaido, as well as accommodation there, see the relevant entries in the "Town by town" section—i.e. Abashiri, Akan, Hakodate, Noboribetsu, Sapporo, Shiraoi. See also the "Itineraries" chapter in the "Japanese Journey" section of the book.

honshu (island of)

■ Honshu was the cradle of Japanese civilisation. Bounded on one side by the Sea of Japan, on the other by the Pacific Ocean, it forms a great arc 1,200 kilometres long and only 250 kilometres across at its widest point. It is by far the largest of the Japanese islands, with a population of 83 million. Honshu can be divided into six regions, from north to south. Tohoku, in the north, is the least developed. It is a thickly-forested mountain area where the winters are severe and dry on the Pacific side and very snowy on the Sea of Japan. It lives mainly from agriculture; rice is the principal crop. An industrial zone is being developed at Akita on the shores of the Sea of Japan. The Pacific coastline is deeply indented, with many offshore islands; it attracts many visitors, particularly to Matsushima which is one of the three great traditional tourist areas of Japan.

The Kanto region is highly industrialised; a narrow coastal plain bounded by hills rising up towards Mount Fuji. It enjoys a continental climate. It is very densely populated, with 30 million inhabitants, including those of the capital, Tokyo.

A 250 kilometre strip across the island, stretching from Tokyo to Nagoya on the Pacific side and including the Japan Alps, forms the Chubu region. This is a very varied province: in the Izu Peninsula and the Bay of Ise a rocky coastline is interspersed with sandy beaches, rice grows on the plains and tea is cultivated on the hill-sides. There is extensive industrial development around Nagoya. The Japan Alps rise to a considerable height in the centre of the island. On the eastern coastal plain there are ricefields again, extending often down to the dune-fringed shoreline on the Sea of Japan.

The Kinki or Kansai region, with its two ancient capitals at Nara and Kyoto, constitutes the historic heart of Japan. This is another varied province, including the industrial areas on the Inland Sea, the great cities of Kobe and Osaka, countryside where rice is grown, and gently rolling hills near Kyoto as well as some quite respectable mountains. In winter there are often heavy snowfalls on the shores of the Sea of Japan whilst the climate on the Inland Sea is exceptionally mild.

Chugoku comprises the whole of the southern part of Honshu. The coastal strip linking Kinki province with the Island of Kyushu becomes more heavily industrialised as you move south from Okayama through Himeji to Hiroshima; the hinterland however, is still very traditionally agricultural.

For information about visits and tours in Honshu, as well as accommodation, see the various entries for towns and sites in Honshu in the Town by Town section.

*This tiny island crowned with pines
is typical of the scenery on the Sea of Japan,
it's a miniature landscape garden in itself.*

ise shima

■ The Ise Shima Peninsula, which has been declared a National Park, covers an area of over 52,000 hectares. Its wooded mountains slope gently down to an indented coastline whose pretty bays are scattered with hosts of little islands. Ise Shima is very popular with Japanese visitors who come on pilgrimages to the shrines of Ise and make trips up the coast, especially to Toba, a centre of the cultured pearl industry.

Ise Jingu shrines

Among the hundred and twenty or so Shinto shrines in Ise the two most venerated are those of Gegu (the Outer Shrine), sacred to the Goddess of Grain and Harvest (Toyouke bime no kami), and Naigu (the Inner Shrine), sacred to the Sun Goddess (Amaterasu o mikami). These Great Shrines of Ise (Ise Jingu) stand in a splendid park, amid groves of cedars and pines through which wind cool clear streams. Some visitors have found that there is little to see at Ise; yet it is a magical place, the very heart of that old Japan still faithful to ancient beliefs in spirit beings, the kami, who are felt to be present in trees and plants, rocks and watercourses.

The shrines of Naigu and Gegu stand a few kilometres apart, deep in the woods. They are small buildings, wooden, neither painted nor lacquered, standing on piles, roofed with thatch, and surrounded by a staked palisade. They are in fact copies of houses of the Yayoi period (see "Japan through the centuries", and "The arts in Japan"). "Alternating between two similar sites, this most sacred of shrines is periodically transferred from one to the other, i.e. actually dismantled and reconstructed. Thus there is an occupied site and a vacant one; vacant that is except for a low pillar, symbol of the unceasing cycle of life and of the Shinto belief in the perpetual hope of resurrection." (D. and V. Elisseeff.) The temple buildings are demolished and rebuilt every twenty years. The wood is used to make objects for sale to the pilgrim visitors.

The Gegu Jingu is reached through a great "torii" and down a long avenue. Ordinary mortals are not allowed in beyond the second enclosure. Since the 4th century the Naigu Jingu has housed a sacred bronze mirror, one of the three Imperial Attributes, originally given by the Goddess Amaterasu (see: *Religion*) to her grandson, Ninigi no Mikoto, ancestor of the Emperor Jimmu, as a symbol of her own image. The avenue that leads to the shrine is lined with various pavilions, places of purification for priests and emperors, a dance pavilion, and the sacred stable. Steps lead down to the banks of the Isuzu gawa, where the pilgrims purify themselves in the crystal-clear waters. After passing still more pavilions, where sake is sold and food offerings are prepared, at the top of a few steps the simple building housing the sacred mirror is reached. The Emperor and the High Priest alone are permitted to enter the final shrine. The ordinary pilgrims stop here, make their offerings, and pause to meditate.

Toba, the pearls realm

His duties to the gods performed, the modern Japanese pilgrim will now proceed to relax...

At Futami, some twenty kilometres from Ise, near the coast, stand two rocks, the larger surmounted by a small "torii": they represent husband and wife, the eternal couple, Izanagi and Izanami (see: "*Religion*"). These two rocks are joined by a rope of straw which is renewed each year on the 5th January. If it should break during the year it is accounted an evil omen.

Toba, a small island 16 kilometres from Ise, is the centre of the cultured pearl industry and the Mikimoto family seat. It was Kokichi Mikimoto who, in 1893, after years of patient research, first succeeded in producing these pearls. Since the Second World War the cultured pearl indus-

kagoshima

try has expanded enormously. At Toba, on a little island rather swamped now by mass tourism, there is a museum where one can see both a film and a demonstration of the production of cultured pearls. Finally there is the spectacle of women "ama-san", diving for pearls from a boat. There are of course many shops to tempt the visitor...

From Toba it is possible to take ferry-boats to the little islands of Hyuga jima, Suga jima, and Kami jima (where the Oceanographic Museum is worth a visit) which are all three devoted to the pearl industry.

ISE SHIMA

Location and access: Island of Honshu (S.E.). National Park covering 52,036 hectares, one of the most visited in all Japan. Organised one or two-day tours from Osaka and Kyoto. Details from JTB offices. Kyoto 160 km, Osaka 140 km. From Osaka (Namba Station) JNR trains run to Toba (2 hrs). Post code 517. Telephone code 05992.

Accommodation: Toba: *Toba International Hotel****, 1-23-1 Mondo Misaki, 94 rooms + 30 trad. style. Magnificent position on the peninsula with beautiful view of Pearl Island. Ryokan in Toba: *Fujita Toba Kawakien****, 1061 Arashima cho, tel. 5-3251, 122 rooms. *Hotel Taiikie****, Ohama cho, tel. 5-5111, 60 rooms. *Sempokaku***, 2-12-24, Toba cho, tel. 5-3151, 46 rooms.

Festivals: Ise Shrines: 4-17 Feb., 14 May, 15-17 June, 15-17 Oct., 23 Nov., 15-17 Dec.

Souvenirs: Cultured pearls.

■ The first thing to do in Kagoshima is to climb up to the look-out point on Shiro Yama Hill, right in the centre of the city. From here there is a fine view out over Kagoshima, which is separated by a narrow inlet from the highly photogenic Sakurajima volcano with a plume of white smoke rising from its crater. Some people compare Kagoshima with Naples—there are the same blue seas, mild climate, and beautiful sunsets —though others point out that there is no real equivalent to its colourful Italian bustle here in Kyushu.

The second essential excursion, if the weather is fine (if it's raining or cloudy, the sulphurous fumes and —sometimes—falling ash may not be to the taste of the sensitive visitor!), is to make a trip by ferry to what used to be the island of Sakurajima (ferry from the Sakurajima jetty), dominated by its imposing conical volcano which erupted in 1914 and formed the bank of lava which now links it to the mainland. There are always taxis waiting to take visitors from the ferry berth up to the look-out points at Yunohira and Arimura —both high up in the lava fields and close to the craters themselves. In the mornings there are organised bus tours available. There is also a pleasant trip by local bus to the nearby fishing villages or to Furusato, a popular spa situated at the foot of the volcano. The fertile soil hereabouts produces fine fruit and vegetables —particularly giant radishes, sometimes as much as a metre in girth.

From the late 12th century, for almost the next eight hundred years, Kagoshima was the feudal seat of the Shimazu family. St. Francis Xavier spent a year here when he arrived in Japan. In 1886, the Shimazu, with the samurai chief Saigo Takamori at their head, rebelled against the Shogunate. Saigo negotiated the surrender of the Shogun's army and was treated as a national hero. But he wanted to end all foreign influence and fomented a further revolt which was suppressed; he committed "seppuku" (suicide by ritual disembowelling) in 1877.

Kagoshima is a major industrial

kamakura

centre for southern Kyushu, its factories are concentrated around the Kiire-Cho oil complex to the south of the city. If you have time it is well worth visiting the ruins of the Tsurumaru castle, taking a stroll through the lovely Iso koen Park—former home of the lords of Shimazu, and having a look at the statue of Saigo, near the Terukuni jinja temple. Anyone interested in pottery should visit Ijuin (18 km by train) which is a centre for the manufacture and sale of Satsuma ware, cream porcelain with crackle glaze, usually decorated with coloured floral motifs. A full day is needed for even a rapid visit to Kagoshima.

KAGOSHIMA

Location and access: Island of Kyushu. Population 500,000. Capital of Kagoshima Prefecture (pop. 1,740,000). Distances by train: Tokyo 1,463 km, Fukuoka 313 km, Miyazaki 127 km. Airport (45 km north) with direct flights to Tokyo (1 hr 50), Osaka (1 hr 05), Nagoya (1 hr 15), Nagasaki (45 mins), Okinawa (1 hr 10). Port with services to Osaka and all the islands of Japan. Post code 890/892. Telephone code 0992.

Travel agency: JTB, Sengoku cho, tel. 22-8155.

Accommodation: Shiroyama Kanko Hotel***, 41-2 Shinshoin cho, tel. 24-2211; at the top of Mount Shiroyama, 10 minutes by car from the station. 577 rooms + 44 trad. style. 1 hr by car from the airport. *Kagoshima Sun Royal Hotel***, 8-10 Yojiro 1-chome, tel. 53-2020. 321 rooms + 16 trad. style. Conference room seating 3,000; in the city centre with a panoramic view of the Bay of Kinko. *Kagoshima Hyashida Hotel***, 12-22 Higashi-Sengoku cho, tel. 24-4111; telex 7827-07. 200 rooms, western style only; in the city centre; panoramic restaurant serving Japanese, Chinese and Western cuisine. Ryokan: *Kagoshima Kokusai Kakumeikan****, Shiroyama cho, tel. 23-2241. 53 rooms. *Kagoshima Daiichi***, 1-4 Takashi machi, tel. 55-0256. 48 rooms. Plus about twenty more inns of all categories.

Souvenirs: Satsuma ware.

■ Kamakura nowadays lies within the Tokyo commuter belt. The city is situated in an attractive valley, surrounded by lush green hills; it is also a very busy seaside resort and in summer its beaches are crowded. Yet, in the city itself, every stone and monument are redolent of a feudal past.

It was in the year 1192 that Minamoto Yoritomo, having finally defeated his enemies, the Taira (see: *Japan through the centuries*), and received from the Emperor the title of "Sei i Tai Shogun" (chief general against the barbarians), set up his "government from a tent" (bakufu) on his lands at Kamakura. Yoritomo was the first in the long line of shoguns who were to rule Japan until the Meiji Restoration; they had their capital here at Kamakura for more than a hundred and fifty years. For fear that other members of his family might oust him and seize power themselves Yoritomo had them all put to death.

When he finally died his son was unable, alone, to withstand the claims of the Hojo family who ruled in effect for a century, with the title of "shikken" or "regent of the shogun". It was during this period that the Mongols launched their attack and saw their fleet dispersed and defeated due to the "Kamikaze" (wind of the gods). This victory was fortunate for the government at Kamakura. Yet discontents arose, the nobles complained that they had not been duly rewarded for their services in the campaign against Qubilay Khan.

After a short exile, the Emperor Go Daigo returned to Kyoto, with the support of the general in command of the Bakufu army, Ashikag Takauji. He in turn was named as shogun by the Emperor Komyo and established himself in Kyoto, whilst the Emperor Go Daigo set up his court at Yoshino, near Nara. Ashikaga Takauji's son and his successors remained in Kamakura, engaged in constant disputes with their Uesugi ministers and thus diminished the power of the regime there. When the Tokugawa made Edo (Tokyo) their new capital it was the end of Kamakura which had been at the centre of Japanese life from the end of the twelfth century until the middle of the fourteenth.

This had been the glorious period of Japanese feudalism, a time of continual conflict when the whole of life was dominated by the military spirit and the idea of heroic vaour. "The exquisite refinement of the courtiers and nobility of Kyoto had been succeeded by a heroic cult of purity and austerity, of truth and toughness." (F. Maraini.) The philosophy of "Zen", with its insistence on personal effort to attain enlightenment appealed to the men of this period who were striving for a sort of self-control; Kamakura became a "Zen" centre and the poet Eisai or Yosai became its best-known teacher. The "Zen" temples were the most beautiful in all Kamakura.

At that period the city was strung out between the hills and the sea. It had a population of more than 200,000 and was several times sacked, burned or devastated by tidal waves, during the fifteenth century. Only a few temples and wooden palaces survive. The Great Buddha is often the only one of Kamakura's art treasures shown to visitors on organised tours. Yet there are many temples well worth visiting.

The Dai Butsu or Great Buddha

The Dai Butsu or Great Buddha dates from the middle of the 13th century. It was built by Maseko, wife of Minamoto Yoritomo. The statue, over 11 metres high, is made of plates of cast bronze, rivetted together from inside and then chiselled in situ. It represents the god Amida, seated crosslegged and leaning forward slightly. "He appears to be dreaming the purest, most peaceful of dreams. His massive head and the sweet, meditative expression on his face suggest the idea of infinite intelligence, free from all illusion, of infinite goodness, purified of all egoism, of infinite happiness, born of the renunciation of all desire." (Pierre Loti.) Visitors can climb up inside the statue. The Great Buddha sits in meditation in the midst of a peaceful garden—usually thronged with visitors unfortunately! It is some distance away from the temples. If time is short a taxi is the best way to reach them.

Wars, violence and dance

Tsukugaoka Hachiman gu, the temple of the god of war, stands on a wooded hillside. Founded by an ancestor of Yoritomo, Minamoto Yoriyoshi, in 1063, it was rebuilt in the 16th century and then again in the 19th. This temple is reached by taking the street called Wakamiya, where there are three great stone "torii" (gateways). At the foot of the monumental staircase leading to Kamino Miza (the main shrine), which is painted red and decorated with sculptured animals and numerous birds, stands the "Maidono" or Dance Pavilion. This was where Shizuko, mistress of Minamoto Yoshitsune, brother of Yoritomo, was obliged, while pregnant, to perform an elaborate dance; some say this was so that she would give birth prematurely, others that it was a trick to make her reveal her lover's hiding place. Close by, a ginko tree, said to be more than 1,000 years old, casts a shadow where Minamoto Sanetomo was stabbed to death by his nephew Kugyo, high priest of the temple. The shrine comprises a gallery and the principal hall or "Honden"; it contains various relics including armour that belonged to Yoritomo.

The Museum, Kamakura Kokuho kan, was designed by a pupil of Le Corbusier; it houses a fine collection of painting and sculpture of the Kamakura and Muromachi periods. "In Japanese sculpture the Kamakura period was characterised by a kind of tense, nervous grandeur, a sort of realism taken almost to baroque—a peak but yet an end. The statues of the period reflect the times that gave them birth. Here are no serene victorious gods, rather venerable human

*The Great Buddha of Kamakura, lost in meditation,
free from all illusions and desires,
pays no attention to the crowds at his feet.
Perhaps he is dreaming of distant feudal times...*

figures marked by suffering. What were needed were examples to instruct the humble faithful masses." (Elisseeff.)

Amongst the museum's collection the graceful statue of Suigetsu Kannon, the wooden figure of Shoko o (1250) and that of Minamoto Yoriyoshi are particularly worth noting. The paintings on display are "emakimono" or painted scrolls, depicting the everyday life of the people.

Between the Hachiman Shrine and the Kamakura gu is the tomb of Yoritomo, a simple pile of stones close to a little pagoda.

Sugimoto dera, founded in 734, is the oldest temple in Kamakura. It contains three wooden statues of the Eleven-Headed Kannon.

In what used to be the main street of Kamakura, Komachi koji, stands a monument to the memory of the priest Nichiren; he was famous around the year 1253 for his vigorous preaching, highly critical of the government of the time (see: *Religion*). He was exiled several times. His ashes lie in the neighbouring Hongaku-ji. Myohon-ji contains a fine stone figure of this priest (13th cent.).

The great Zen priest and preacher, Eisai, introduced the cultivation of tea into Japanese, claiming that it helped to clear the mind and was an aid to meditation. The temple of Jufuku-ji, built in 1200 by the wife of Minamoto Yoritomo, contains a fine wooden statue of Eisai, as well as a Jizo figure with striking eyes made of jade.

Kencho-ji Temple (2 km. north of the city) was built in the 13th century, by a "Hojo", for a Chinese priest; it was rebuilt in the 16th century, after a fire. Standing in a fine setting of cedar trees it is one of largest temples of Kamakura. Passing through the "Kara mon" (doorway) we enter the "Hon do" which contains fine paintings and sculptures.

North of Kamakura, again surrounded by splendid trees, stands the great Zen temple of Engaku-ji. After the great two-storied gateway, San Mon, we come to the belfry which contains a massive bronze bell, 2.5 metres high, cast in 1301. The Shari den or "relic pavilion" contains what is claimed to be a tooth of the Buddha. The reliquary dates from 1285 and is an example of Zen design, strongly influenced by Chinese architecture of the Sung Period —as its construction and high elegant roof clearly indicate.

"The Honourable Picnic"

After seeing all these lovely things in Kamakura a little rest and relaxation is probably what is required. For this there are pleasant beaches nearby (Zaimouza for instance) or the little island of Enoshima. Anyone who expects Enoshima to be still as Thomas Raucat depicted it in "The Honourable Picnic" is likely to be disappointed. The island is now joined to the mainland by a wide roadway and many large buildings have rather spoiled its charm; but there are plenty of nice little shops and inns where good shellfish are served. An amusement park and marina are less pleasant features. On Sundays the Benten Shrine and Grotto (said to have been the lair of a child-eating dragon) are crowded with sightseers. The Goddess Benten, the only woman among the Seven Gods of Happiness, is said to have created the island, flown down there one day, fought the dragon, defeated him and then married him—thus putting an end to his crimes!

KAMAKURA

Location and access: Island of Honshu. Tokyo, 45 km, Yokahama 21 km. Seaside resort and residential suburb for both cities. Tours readily available from Tokyo.

Travel agency: JTB, Sengoku cho, tel. 22-8155.

Accommodation: Visitors to Kamakura are recommended to stay at an hotel in Tokyo and take an organised tour from there to Kamakura.

Festivals: Kamakura Festival, 7-14 April; Festival at the Tsurugaoka Shrine 16 September, mounted archers.

kanazawa

■ Kenroku-en, one of the most splendid landscape gardens in Japan, is the pride of the city of Kanazawa and attracts thousands of visitors every year. The city itself, built between two rivers, 10 kilometres inland from the Sea of Japan, retains much of its feudal splendour. It is still a city of craftsmen: silk-weavers, lacquerers and wood-carvers (who work mainly in paulonia wood).

In the late 16th century Kanazawa owed its prosperity to the powerful Maeda family, related to the shogun Tokugawa Ieyasu. During the Edo Period (1603-1868) these wealthy daimyo—they received more tribute money from their vassals than any other lords—built the castle and established the prosperity of the city.

Only one gateway remains of the Maeda Castle, built on a hill and destroyed by fire in 1881. Its site is now occupied by the University of Kanazawa.

The Gardens of Kenroku-en are situated some 3 kilometres from the city (take the bus from the station). They were laid out for the Maeda family in the eighteenth century and have been open to the public since the beginning of the Meiji Period in 1868. They contain miniature mountains and two large pools, while little waterfalls and streams lend an atmosphere of life and freshness. It is possible to identify, here at Kenroku-en, all the elements of the "classic" Japanese landscape garden (see: "The Art of the Garden"). Near one of the pools stands an eighteenth-century tea house, designed by the artist Enshu. The flowering cherries (kiku-zakura), especially a rare type known as "chrysanthemum cherries", and the marvellous irises, attract enormous crowds of admirers to Kenroku-en in springtime.

KANAZAWA

Location and access: Island of Honshu, 10 km inland from the Sea of Japan. Pop. 700,000. Tokyo 522 km, Gifu 226 km, Komatsu Airport: 35 km.
Post code 920. Telephone code 072.

Travel agency: JTB, Fukoku Seimei Bldg., Shimozutsumi cho (tel. 61-6171).
(See other notes p. 121)

kobe

■ Kobe is just over a hundred years old. Its harbour was built at the beginning of the Meiji Period (1868) when the increase of trade through nearby Osaka created the need for a deep-water port.

To get some idea of the extent of Kobe it is a good idea to climb the Rokko san, the dragon-shaped mountain that overlooks the city. (There is a bus from the Sannomiya Station to the funicular, or you can take a taxi.)

Kobe is the sixth largest city in Japan, with 1,350,000 inhabitants. Built on a narrow coastal plain two kilometres wide, it is gradually expanding, thanks to massive land reclamation on the sea-ward side and a tunnel under Mount Rokko; it needs all the land it can get, for its expanding industrial and residential areas. The view from the Harbour Tower gives an idea of the size of the port. In the foreground are clusters of strange covered shallops that look like gigantic wood-lice; the larger vessels are moored further out. The Harbour and Naval Museum, on the first floor of the tower, is worth a visit to see its working models of harbour installations.

A trip round the harbour by motor boat (departures from Naka Pier at the foot of the Harbour Tower) gives a good close view of the Kawasaki and Mitsubishi shipyards, of submarines, tankers and cargo vessels in dry-dock, gigantic cranes, and of ships flying all the flags of the world. Japan's leading commercial port, Kobe handles 34% of all the country's exports (steel, textiles, machinery, metal and chemical products), as well as 16% of its imports (iron ore, cereals, crude oil, raw cotton). An artificial island has been created in the harbour, with earth and rocks brought down from the mountains; it is linked to the mainland by the Kobe-o-bashi, a two-storeyed bridge.

As well as industrial areas, Kobe also has picturesque parts such as Sakemachi dori and Motomachi dori. Hidden among trees at the foot of the mountain stand the red pavilions of the Ikuta inja Shrine, founded in the third century.

*The world's leading fishing fleet
brings in a rich harvest of fish of every kind;
dried octopus is a much esteemed delicacy.*

kotohira

The Museum of Namban Art (Namban = "Barbarians from the South"), (at 4-35-3 kumochi cho 1-chome, Fukiai ku, open 9-4.30) contains an ancient map of the world and two marvellous painted screens of the Kano period (16th-17th cent.); they depict little vessels sailing on a midnightblue sea, against an old-gold background Portuguese figures move, dressed in exotic period costume; people are being carried in litters, followed by elephants, servants carry gifts—birds, lions in cages. Elongated, pale-faced missionaries advance in a dignified manner...

Kobe is the departure point for trips to the islands of the Inland Sea, including Awaji shima.

The spa of Arima Onsen (24 km) is easily accessible by cable-car from Mount Rokko. A full day should be allowed for Kobe—the harbour tour in the morning and the Museum and Rokko san in the afternoon.

KOBE

Location and access: Island of Honshu, pop. 1,500,000. Capital of Hyogo ken Prefecture (pop. 4,920,000). Distances by train: Osaka 34 km, Kyoto 76 km, Tokyo 565 km (3 ½ hrs). Post code 650/657. Telephone code 078.

Travel agency: JTB, Sannomiya eki, Nunebiki cho, 4-chome, tel. 231-4701.

Accommodation: New Port Hotel***, 7-3-3 Hamabe dori, Fukiai ku, tel. 231-4171. 205 rooms. City centre. Panoramic views from revolving restaurant. *Oriental Hotel**** 25 Kyomachi, Ikuta-ku, tel. 331-8111. 204 rooms; new building in city centre with harbour view. *Rokko Oriental Hotel****, 1878, Nishi Taniyama, Rokkosan cho, Nada ku. 45 rooms + 15 trad. style, tel. 891-0333; at the top of Mount Rokko, splendid views over the harbour and the Bay of Osaka. *Kobe International Hotel***, 1-68-chome, Gokodori, Fukiai-ku, tel. 221-8051, 45 rooms + 3 trad. style; city centre. Ryokan: *Hotel Zentain****, 4-43-1, Shimo Yamate dori, Ikuta ku, tel. 391-3838. *Rokko Soky Villa****, Nishi Taniyama, Sumiyoshi cho, Higashinada-ku, tel. 891-0140, 36 rooms. *Hotel Sannomiya Central***, 4-2-1 Nunobiki-cho, Fukiai-ku, tel. 241-5031, 20 rooms + another ten ryokan, three, two and one star.

Restaurants: Blanc du Blanc** (Japanese and Western cuisine) 7th Floor, Shinei Bldg., 1-77, Kyo machi, Ikuta ku, tel. 321-1455. *Coral Kitano*** (Western cuisine) 31-64, Kitano cho, Ikuta ku, tel. 231-2251.

Youth Hostel: Tarumi Kaigan, 5-58 Kaigan dori, tel. 707-2133.

■ The little town of Kotohira, with its charming little houses—often converted now into restaurants and small shops—is famous as a place of pilgrimage, to the shrine of Kotohira gu or Kompira san. The seven hundred steps leading up the Zozu san are always crowded with pilgrims, on their way to the various temples to make their offerings to Okuninushi no Mikoto, a descendant of Izumo, and to the god Kompira —whose origins are shrouded in mystery but who is venerated as a protector of sailors and travellers. A few further flights of steps lead up to the "Dai mon" and to "Sho in".

Yet more—very steep—steps bring us to the Tea House and the Hall of the Rising Sun. Finally there is a long avenue, lined with stone lanterns, then a path through the woods, to reach the "Okuno in"; from here there is a fine view, extending as far as the Inland Sea. Tired pilgrims can avail themselves of "kago" (a kind of sedan chair, carried by porters) to make the ascent in comfort. There are many stalls and kiosks selling all kinds of souvenirs; shopkeepers and restaurateurs often give walking sticks and umbrellas to customers who have made the pilgrimage on foot!

On the way down it is pleasant to pause in a modest cafe for a while, sit on a wooden bench in the cool, and sip a bowl of slightly bitter, herbal-tasting tea...

KOTOHIRA

Location and access: Island of Shikoku. 24 km from Takamatsu (2 hrs by train). A little town, nestling beneath the shrine of Kompira san. Post code 766. Telephone code 08-777.

Accommodation: It is advisable to stay at Takamatsu, though there are 3 de luxe ryokan at Kotohira: *Kotohira Kadan****, Kotohira cho, Nakatado gun, tel. 5-3232. 43 rooms. *Kotohira Grand Hotel****, Nakatado gun, tel. 5-3218. 43 rooms. *Kotohira Korusai Hotel Yachiyo***, tel. 5-3261. 50 rooms. All on the avenue leading to the temple.

Souvenirs: Little wooden boats containing pilgrim figures.

koya san

Kobo Daishi founded the first monastery of Koya san up here in the mountains at the beginning of the 9th century. It soon became the headquarters of the Shingon Sect of Buddhism. Protected by the Emperor and the Court, the monastery rapidly became extremely prosperous; it is said that during the Middle Ages this monastic complex comprised 1,500 buildings housing some 90,000 monks. Over the centuries the monasteries were ravaged by fire and the existing buildings are mainly 19th-century reconstructions. There are about a hundred temples in the area today, much visited by pilgrims to whom the Buddhist monks offer frugal but friendly hospitality (they follow a vegetarian diet).

Koya san is magnificently situated amid mountains and forests. The Abbot Superior resides in the principal temple, the Kongobu ji. Its sacred enclosure or "garan" is dominated by a great pagoda. A nearby building, the Fudo do, houses fine statues of "guardians of the temple". The Museum or "rehoka" shows a changing selection of the monastery's 25,000 treasures. Amongst those on permanent display are painted scrolls, one of which depicts a great serene figure of the Buddha surrounded by 25 "boddhisattvas" smiling in expressions of celestial beatitude. On the outskirts of the village of Koya san, a paved pathway leads to the mausoleum of Kobo Daishi. It crosses the great necropolis of Koya san where, under the trees, stand thousands of monuments to the great families of Japan—the Tokugawa, the Asano and the Date.

(See notes p. 121)

THE KAGO OR PALANQUIN

■ "*Kago, strictly speaking, means basket. This is what it really is. Imagine an oblong trunk, with two handles, big enough to hold a human body folded into three; a long thick shaft is passed through the two handles and the whole thing is covered with a roof of wicker-work. You slide about in it on your side, your knees drawn up under your chin, carefully moving your head to avoid the shaft that holds the whole contraption. Unless you have practised this sort of contortion from an early age you soon find yourself in an extremely uncomfortable, and highly inelegant, posture. Two strong porters take hold of the ends of the shaft, lift it onto their shoulders and stride along, using bamboo canes to help them. If it should come on to rain they will envelop you and the whole contrivance in a covering of oiled paper. Thus reduced to the status of a parcel you will become unconscious of your surroundings, except to notice that you plunge forward headlong when your porters are going downhill, and backward when they climb.*"

GEORGES BOUSQUET
"Japan Today" (published in 1870)

kurashiki

- Close to the industrial centres of Mizushima and Tamashima, the old town of Kurashiki has retained both life and charm. The town's prosperity in the old days was founded on the rice trade; its name, the diminutive form of Funakurayashiki, has the same character as that for grain and sake stores. The Tokugawa shoguns encouraged the rice trade and the merchants of the period built themselves fine houses alongside their granaries—they can still be seen in what is now known as the Museum District.

Charming old wooden houses line the banks of Kurashaki gawa, up which the riceboats used to come; it has now been canalised and planted with willows. A curious Neo-Classical building houses the Ohara Art Museum, the famous collection formed by Ohara Magosaburo a benefactor and restorer of the town. Western Art is represented by El Greco (The Annunciation), Renoir (Jeune Femme à sa Toilette), Monet (Waterlilies), Gaugin (Le jardin parfumé), as well as paintings by Pissaro, Picasso, Cézanne, Marquet, Matisse, Utrillo and Rouault, and sculptures by Rodin and Bourdelle. A whole floor is devoted to contemporary Japanese artists. One room contains archaeological finds from Asia and the Near East. Another two rooms contain collections of wood engravings and printed fabrics. Close by, the Folkcraft Museum, housed in four former rice granaries, contains a very interesting collection of fine Japanese craftwork, classified according to the regions where they were made. The Toy Museum has toys from all over the world, but its most important and interesting section is its collection of Japanese toys. Finally, in another granary, opposite a little bridge, there is the Archaeological Museum with objects from the Kurashiki region as well as others from China and even some from South America.

Kurashiki amply repays a full day's visit and it is well worth making a trip to the nearby little fishing port of Shimotsu.

KURASHIKI

Location and access: Island of Honshu, pop. 400,000, in Okayama Prefecture (pop. 1,900,000). Distances by train: Tokyo 694 km, Hiroshima 154 km, Osaka 206 km, Shinkansen from Shin-Kurashiki Station: Tokyo 5 hrs 15 min., Fukuoka 3 hrs. Post code 711; telephone code 0864.

Accommodation: Kurashiki Kokusai Hotel****, 1-1-44, Chuo, tel. 22-5141, 70 rooms; town centre, close to Museum district. *Mitzushima Kokusai Hotel****, 4-20 Mitzushima Aoba cho, tel. 44-4321, 75 rooms. 20 min. by car from the station. Near the industrial district. 1½ hrs by car from Okayama Airport.
Ryokan: *Shimoden Hotel****, Obatake, tel. 79-7111, 137 rooms. *Washu Grand Hotel****, Shimotsui, tel. 79-8222, 78 rooms.

Youth Hostel: Kurashiki, Minamiura, Sambonmatsu, Mukaiyama, tel. 22-7355.

Souvenirs: Pottery, plaited straw bags from Igusa, wooden dolls, bamboo objects.

KOYA SAN (p. 120)

Location and access: Koya is the official name of the district where the monasteries of Koya san—the great centres of Japanese Buddhism—are situated. Distances by train: Tokyo 580 km, Osaka 60 km. Accessible from Osaka by the Nankai Railroad (trains from Namba Station); take the train to the terminus and then take the funicular to Koya san. Organised bus tours are available from Kyoto and Osaka, details available at JTB offices.

Accommodation: There are neither hotels nor ryokan at Koya san; the monasteries offer hospitality to visitors, it is an experience worth trying.

KANAZAWA (end of p. 117)

Accommodation: Kanazawa Miyako Hotel***, 6-10 Konohana cho, tel. 31-2202. 90 rooms; opposite the station. *Kanazawa New Grand Hotel****, 1-50 Takaoka machi, tel. 33-1311. 120 rooms; 5 min. by taxi from the station, near Kenroku en Park. *Kanazawa Sky Hotel*, 15-1 Masashi machi, tel. 33-2233, 150 rooms; city centre, with a fine view of the castle.
Ryokan: *Mitakeya****, 17-1 Shimo-Tsutsumi cho, tel. 31-1177.

Restaurants: Kaga Sekitei**, 1-9-23 Hirosaka (Japanese cuisine), tel. 31-2208. *Tsubajin Wako,* 1 honda cho (Western cuisine), tel. 63-5525.

Festivals: Hyakuman Goku Festival in mid-June.

Souvenirs: Lacquered objects and painted silks.

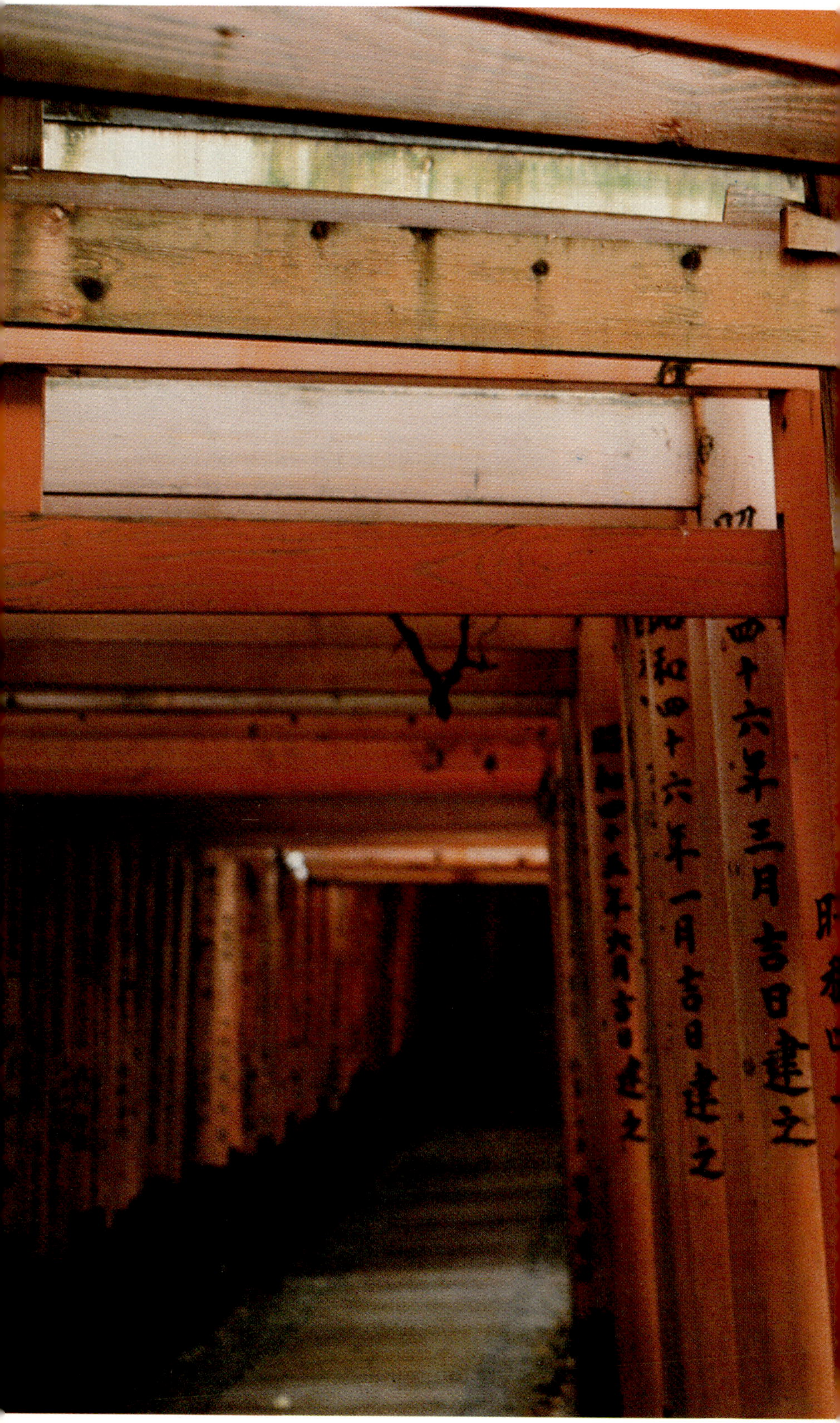
昭和四十六年一月吉日建之
四十六年三月吉日建之

*Ten thousand pillars, each engraved with the name of its donor,
are set aflame by the setting sun
and become a glowing avenue leading to
the popular Inari Shrine at Kyoto.*

kyoto

■ With its 1,500,000 inhabitants Kyoto is the fourth-largest city in Japan today. Thanks to the intervention of the French Orientalist, Elisseeff, Kyoto was not bombed at all during the Second World War. The city lies several kilometres away from the great industrial areas of Japan and its own factories are concentrated in an area south of the railway station, so it remains miraculously intact and free from industrial pollution. Ten million visitors come here every year. Kyoto has often been compared to Florence, both cities being in some sense the artistic capitals of their respective countries, rich in museums, temples or churches. The Italian writer Maraini considers that there is even a physical resemblance, they both "extend like lakes across the plain, between mountains and hills". But "Florence is quintessentially Western, beautiful, a splendid goblet that one can drain in one draught". Kyoto is Eastern, hides her beauty, yields her secrets with reluctance. Kyoto has a magic power, "a soul, a message, a unique personality of her own" (Maraini).

The Circle of Oriental Delights

"The city can be divided roughly into two parts: a belt of wooded hills, scattered with lakes, temples, villas, gardens, hermitages and monasteries, a circle of oriental delights; in the centre an urban nucleus, with all the features of a modern city, its famous "geisha" districts, theatres, restaurants, antique shops, busy streets, luxurious houses, an island of earthly delights." (Maraini.)

Kyoto through the centuries

In 784, to escape the influence of the Buddhist monasteries, the Emperor Kwammu decided to leave the city of Nara which had been the Imperial headquarters since the 8th century. In 793 he chose the site of the new capital, encircled by mountains, on the banks of the river Komo gawa. The Emperor named the city Heian jo, the "City of Peace"; later it was called Miyako, "capital", and finally Kyoto, the "Capital City"; it was to remain the capital of Japan for a thousand years.

Kyoto was built on a site five kilometres square, to a geometric plan deriving from the Chinese city of Tch'ang an, the Tang capital. An avenue 85 metres wide linked it with the "Southern Gate". There were nine "diyo" or straight streets, bearing numbers rather than names, running from East to West, crossed at right angles by other streets. The streets of Kyoto, bordered by flowering cherries or willows, totalled more than twelve hundred in all... The Imperial Quarter was bounded by a high wall with six gateways; this was where the court nobility and their families lived. The Imperial Palace stood in the centre of this district, a tall red building with a green tiled roof. The "Pavilion of Sumptuous Pleasure", reserved for the celebration of festivals, and the "Celestial Imperial Pavilion", where court ceremonies were held, stood alongside it. The Emperor's own quarters were known as the "Apartment of Sweet Freshness". The pavilions reserved for women and the Empress bore the names of flowers. In the commercial district each street was reserved for a different trade. Lacquerers, goldsmiths, armourers, embroiderers and tea merchants all lived in the quarters reserved for the wealthy members of society.

Kyoto at the time of "The Tale of Genji"

During the 11th and 12th centuries the Emperors Shirakawa, Hori Kwa and Toba succeeded each other as sovereigns. It happened more than once that an Emperor retired or went into temporary "exile" in favour of his successor but in fact continued, secretly, to exercise power. Each had his supporters and this was how the famous conflict between the Minamoto and Taira families (Genji and Heike) arose... The Lady Murasaki Shikibu gave a personal account of the history of this period in her "Tale of Genji", a vast literary work in fifty four books which relates the court life of the lords and ladies of the time.

When Taira Kiyomori was named First Minister, Minamoto Yoritomo was condemned to exile; this proved a short-lived punishment as, after defeating the Taira family once and for all, he took power in 1185 and established Kamakura as the seat of his government "bakufu". In his novel "The Chronicle of the Heike", the contemporary writer Eiji Yoshikawa gives a marvellous picture of Kyoto at this time—of the splendours and miseries of the life of the nobility, as well as the world of the craftsmen and the common people; he describes streets where great ladies rode in carriages drawn by oxen, and the "invasions" of the Buddhist monks from Mount Hiei who descended on the city with their "sacred palanquins", terrorising the inhabitants and demanding land and temples.

The feudal period

The next period, from the 14th to the 16th century, was one of great confusion. Minamoto's successors established themselves in Kyoto and it was they who decided who was to become Emperor. After his exile, the Emperor Go Daigo returned to power, with the support of Ashikaga Takauji who then proceeded to gain control first of Kamakura, then of Kyoto, and eliminated Go Daigo and made Komyo Emperor instead: It was like a Kabuki play come to life. Rival dynasties, "Northern" and "Southern" were set up, but power remained in the hands of the Ashikaga shoguns. Kyoto became the scene of civil wars which devastated the city and spread fear and famine amongst the people. Whole streets were abandoned, buildings were erected without any concern for the overall plan.

The city gradually lost its regular chessboard layout, networks of narrow winding streets grew up. Many people left the centre to live in the hills near the castles. This period of civic disorder lasted until 1573, when Oda Nobunaga seized power in Kyoto and abolished the Ashikago shogunate. The Buddhists had a civilising influence, in their monasteries they cultivated fine feelings, good manners, and intelligent appreciation of nature and the arts. As for the lords themselves they lived lives of luxurious refinement. This was when the marvellous Golden and Silver Pavilions were built, a time when cultured people would meet for the tea ceremony, to compose poetry or to discuss works of art.

The role of Toyotomi Hideyoshi

On the assassination, in Kyoto, of Oda Nobunaga, Toyotomi Hideyoshi took power. At last there was a period of peace; the city developed both economically and artistically. The temples were restored. Hideyoshi patronised painters; great entertainments were held in Nijo Castle and Sen no Rikyu, the great master of the art of tea drinking, decided on the layout of its gardens and tea houses.

Even after the shogunate moved to Edo under the Tokugawa, Kyoto remained the Imperial as well as the artistic and religious capital of Japan. The city was devastated by fire in 1788. At the Meiji Restoration, Edo (Tokyo) became the capital, but the Imperial Coronation took place in Kyoto.

Two days, a week, a month

Kyoto is a somewhat overwhelming experience for a visitor arriving here for the first time. There are over two thousand temples and shrines in this unique, magical place, this city of gold and silver and silk. Yet one's first impression on arriving at the station is of a modern city, seething with people.

If you have only one day in which to see Kyoto the best way to see as much as possible is to take the two JTB tours. The first, morning, tour takes in Nijo Castle, the Golden Pavilion, the Imperial Palace and the craft centre. The afternoon tour includes visits to the Heian Shrine, Sanju sagen do, and Kiyomizu dera. Each tour costs about 2,200 yen.

Visitors with several days or a week or more at their disposal can visit the selection of temples, shrines, villas and other sites given in this guide, in a more leisurely way. A taxi is a good means of seeing the maximum number of things if time is limited. The following is a suggested itinerary:

First morning: Nijo Castle, Kinkaku ji or Golden Pavillon, the stone garden of Ryoan ji.

First afternoon: Silver Pavilion, Kyomizu dera (to see the sunset).

Second morning: Heian Shrine, Sanjusangen do.

Second afternoon: Katsura Villa and, at 2 pm, Nishi Hongan ji (apply for special permission at the temple office).

Third morning: the Shugaku in Rikyu Villa (special permission required).

Third afternoon: Mount Hiei, or Fushimi Inari, or Byodo-in.

Thereafter, if time is available: the shrines and temples outside the city, such as Fushimi Inari and Byodo in; the Daikoku ji monastery, Mount Hiei, the museum, the old quarters of the city, and, especially in summer, the banks of the river Kamo. In the evening it is pleasant to stroll in the picturesque Ponto cho and Gion quarters.

Nijo Castle

From the outside, Nijo Castle (open from 8.45 to 4) looks like a traditional fortified castle: thick walls, look-out posts and a lotus-covered moat enclose vast gardens and

*May is festival month in Kyoto.
For a day the children of the city wear the
glowing costumes of gods and medieval warriors
and take part in nautical jousts at Irashiyama near Kyoto.*

KYOTO 127

buildings that, from the time they were begun by Shogun Tokugawa Iyeyasu in 1603 until the Meiji Restoration in 1868, were the Kyoto residence of the Tokugawa. In 1884 the castle was detached from the Imperial Palace and in 1939 it was given to the city. The finest artists of their time contributed to the decorated and embellishment of its pavilions which are typical of the Momoyama Period.

The entrance to the castle is through the East Gate, then a second, Chinese gate, the Kara mon, which leads into the great courtyard of Nino maru. The Kara mon, surmounted by a wide roof, is decorated with various motifs in wood and metalwork. Its interior walls are decorated with a fighting tiger, lion and dragon. Across the courtyard stand the five pavilions making up the palace itself. They are placed in a slight diagonal, and the corridors that join one to the next have so-called "nightingale floors" whose sqeaks announced the approach of visitors—or possible assassins!

The spacious empty rooms convey an impression of refined luxury with their rare woods, chiselled brass and elaborately carved friezes. The "fusuma" (paper-covered screens) diffuse a gentle light that gives added softness to the paintings on the walls and coffered ceilings—by Kano Tanyu and his school. The "fusuma" in the first rooms are decorated with tigers hunting, fighting and sleeping between clumps of bamboo. The audience chamber, where the shogun used to receive the tributes of the "daimyo" (lords) contains paintings of great pine trees with mossy trunks, against old-gold backgrounds; figures dressed in period costume add to the atmosphere. There are sculptured friezes of peacocks and paeonies, the work of Hadari Jingoro. In other rooms there are storks amongst the pines and flowering cherries—great bouquets of white against a golden background. In the fourth pavilion a heron perches on a fishing boat. One room is decorated with masses of coloured fans, flung apparently at random all over the walls.

Throughout the palace, whenever it's possible to see them, the magnificent gardens seem to come right into the rooms and make a perfect unity with the building itself. "The centuries have combined to embellish this palace. Time has softened the sparkle of things here, so that the interior has become a sort of golden haze. Silence and solitude combine to give the feeling of being in the enchanted dwelling of some princess of an unknown world, a planet quite remote from our own." (Pierre Loti).

The Golden Pavilion or Kinkaku ji

Kinkaku ji, still sometimes called Rokuon ji, (open 9 to 5-30) casts the reflections of its golden facade and roofs onto a little lake whose waters are scattered with lotus plants and whose shores and islets are planted with pine trees. This temple, situated at the foot of Mount Kinugasa, began as the modest home of a state functionary Salonji Kintsune. In 1394 the Shogun Ashikaga Yoshimitsu chose this spot for his retirement, built the Golden Pavilion and created a magnificent garden setting for it. On his death, the villa became the Rokuon Temple which was deliberately burned down by a young monk in 1950. The incident is recounted by Yukio Mishima in his novel: The Golden Pavillon. The existing pavilion is a replica of the previous one; it was built in 1955. The ground floor and first floor are designed as a private house, the second floor as a temple. Walls, ceilings, floors, everything including the balustrades and the phoenix on the roof, is covered with gold lacquer. Today a warm patina is already lending the building an appearance worthy of its former splendour.

The rooms contain Buddhist statutes and some fine painted sliding screens. Kinkaku ji well repays a visit of several hours; strolling through the gardens is a continual delight as every turn seems to open up yet another view of the pavilion. It is a place whose beauty changes with every hour and every season: unreal and dreamlike under snow, basking in the rich autumnal reds of the surrounding woods or taking fresh life in the spring sunshine, the Golden Pavilion never fails to entrance with its perfect beauty.

There are paths through the gardens, leading to numerous little summer houses, to the Sekka tei, a 17th-century tea house; a miniature temple stands not far away.

The stone garden of Ryoan ji

Ryoan ji (open 8 to 5) is the temple of the Zen sect known as "Rinzai". Founded in 1450 it was rebuilt in 1499, then again in 1797 after a fire.

The temple is famous for its garden of stones and sand, first laid out in the 16th century by Soami, a painter and a great master of the tea ceremony. Surrounded by great trees, and bounded on one side by a low wall, on the other by the austere buildings of the temple itself, the garden consists of a rectangular expanse (30 metres by 10) of carefully raked sand on which are arranged fifteen stones, each lying on a thin layer of moss. Some people see them as mountain peaks rising out of the mist or as islands emerging from the sea; others as tigers leaping about on a dry river-bed...

Ryoan ji is a good example of a Zen garden (see: "The Art of the Garden"), designed to be conducive to meditation on the absolute and the insignificance of life. It evokes the unchanging aspects of nature, unaffected by the seasons and the passage of the years. A few metres away, forming part of the temple area itself, is a splendid landscape-garden, mossy and planted with maples. There are waterlilies floating on the little Lake Oshidori, running streams, and tiny waterfalls that gush down out of bamboo pipes. A ravishingly pretty teahouse stands in the greenery nearby.

The Daitoku ji

This monastery (open from 9 to 5) is a complex of more than thirty temples... seven of them are open to visitors. The first was built by the priest Daito Kokushi, at the request of Emperor Go Daigo, during the 14th century. It was soon joined by other buildings which were burned down many times and subsequently rebuilt. The present Daitoku ji consists of a group of temples surrounded by high walls inside which there is a succession of great gateways, pagodas and pavilions shaded by huge trees. The whole complex is a veritable museum. The visitor enters through the two gateways of Chokushi mon and San mon, both displaying Buddhist statues from Korea, on their upper storeys. The great shrine hall of Butsu den contains statues of the Buddha and of the founder of the temple Daito Kokushi. But the places most worth seeing are the Honbo, where there are "fusuma" painted by Kano Tanyu, and the Daisen in. This latter temple is set amidst three gardens, all highly symbolic: a waterfall becomes a river, "life"; this river flows around the islands of the Crane and the Tortoise, "human knowledge", and thence into the Inland Sea, "the ocean of eternity". The halls of the Daisen in were decorated by Soami, Kano Motonobu and Kano Yukinobu, with paintings depicting flowers, birds and the occupations of the countryside.

The Silver Pavilion or Ginkaku ji

Shogun Ashikaga Yoshimasa had the Ginkaku ji (open from 9 am to 4-30 pm) built, in the mid-15th century, as a place for retirement after his abdication. He took the Golden Pavilion as his model, but died before he could have it covered in silver as he intended; later the dwelling was converted into a temple. Soami was charged with the decoration of the apartments; he produced some lovely painted screens. He also designed an enchanting garden. Ashikaga Yoshimasa lived a life of elegant refinement here in his palace, being an ardent practitioner, with Soami, of the tea ceremony.

The garden is reminiscent of a Chinese landscape; it consists of two parts "one, classical, with a lake, the other, a sand and stone garden, is specially designed to evoke the sea and the mountains." (Louis Frederic, "Japan"). The lake is covered with lotus flowers. There are two silvery hills in the garden, one modelled on Mount Fuji, where Yoshima used to go to discuss aesthetics with his friends, whilst from the other he would watch the rising of the moon.

The Kiyomizu dera

Half-hidden among the changing greens of pines and maples, the Kiyomizu Temple clings precariously to its hillside overlooking Kyoto. It was founded in the 8th century by Sakanoe Tamuramoro, and dedicated to the Eleven-faced Kannon at the request of the monk Enchu. The existing buildings erected on the orders of Shogun Tokugawa Iemitsu.

The temple is a very popular place. Crowds of visitors come here every day, by paths through the woods or down the picturesque little lane lined with shops selling pottery and souvenirs, before making the final ascent up flights of steep steps. They pause to admire the shoro, a belfry dating from 1607, and its enormous bell, the three-storey pagoda that houses a statue of Dai Nishi Nyorai, and drink great goblets of purifying water from the fountain, spewed out through a dragon's mouth. Eventually the shoals of schoolchildren and pilgrims disperse to explore the various halls (Hall of the Sutras, Hall of Asakusa...) or to take photos of the Hon do and its terrace overlooking the gorge, which is supported on masses of tall piles. "They say that Japan is a country of theatrical architecture (butaizukuri) or even precipice architecture (kengai zukuri)."

(Elisseeff) "When one finally reaches the verandah at the end, it is a real surprise to find oneself gazing down into a chasm filled with greenery, bamboo woods, marvellously cool, looking curiously foreshortened seen from up above. It's like being on the balcony of some gigantic aerial dwelling." (Pierre Loti "Japonneries d'automne"). A path leads down to the secondary pavilions of Amida do and Okuno in, giving fine views of the Otowa waterfall.

You sometimes see pilgrims standing in prayer right under its freezing waters. In some of the dark halls it is possible to make out pictures of boats that have been hung there as ex-votos. The inhabitants of Kyoto often make the climb up to Kiyomizu dera in order to watch the sunset.

The Heian Shrine

The Heian Shrine (open from 8-30 until 5) was built in 1895 to commemorate the eleven hundredth anniversary of the birth of the city of Heian. It was dedicated, by the Emperor Meiji, to the memory of the Emperor Kammu and to that of his own father, Komei. Beyond the enormous torii gate lies a great sanded courtyard. The main building and its pavilions are Chinese in inspiration, smaller copies of the first Imperial Palace; they are a brilliant red and roofed with green glazed tiles. Behind the shrine stretches a magnificent garden, with winding stone paths, pools and elegant bridges, one of which is covered. Extensively planted with azaleas and "bonzai" pines, the gardens are much visited—especially when the flowering cherries, the azaleas and the purple iris are in bloom.

The 33,333 Kannons of Sanjusangen do

The great hall of Sanjusangen do (open 8 till 5), founded in the 12th century, contains a forest of a thousand and one gilded wooden figures of Kannon, the Goddess of Mercy,

*The Silver Pavilion, at Kyoto,
was an elegant retreat for a 15th-century shogun;
here in this lovely landscape garden
a group of gifted people used to meet
to discuss questions of aesthetics.*

the work of Tankei Unkei and Kozyo. They stand in serried ranks on steps in the hall, which is 120 metres long and divided into thirty-three (hence the name san-ju-san) to symbolise the thirty-three incarnations of the goddess.

In the centre of these rows of statues, all different but possessing the same attributes, stands a great gilded wooden figure of Kannon, with eleven faces and a thousand arms, the work of Tankei himself and completed in 1250. The temple is also known as "The Temple of the 33,333 Kannons", for this is the figure you arrive at, if you count all the faces and images of the goddess on the haloes and foreheads and hands of all the statues!

The corridor behind the main hall contains the statues of the "28 Faithful Followers" of the goddess, works of remarkable realism. Here also is a figure of the third incarnation of the Buddha. In times gone by when poor families sometimes had to sacrifice a baby at birth they used to pray to this third incarnation of the Buddha, so that he would take pity on the infant and guide it to its second incarnation. Statues of this Buddha in temples today are often decorated with a sort of collar and a little red bonnet.

Chion in

The Chion in monastery (open from 9 till 4) is beautifully sited, on the side of a mountain. It was founded by Genchi, a 13th-century priest, near the tomb of the priest Honen; it was frequently rebuilt after being burned down, the present buildings date from 1639. Until the Meiji Restoration it was usual for the Abbot of Chion in to be of royal blood.

Honen was the founder of the Jodo sect (see: "Religion") of which this monastery is the present-day headquarters. The entrance is up great flights of steps, through lofty gateways flanked by marvellous old trees.

The great San Mon Gate is a two-storeyed building. In the temple courtyard stands a pagoda, the great pavilion of Mie do or Hon do, and the Amida do. Mie do contains statues of Honen, Amida (the venerable Buddha of the Jodo sect), as well as of the priest, Genchi. A huge statue of Amida stands in the Amida do. The corridors in the galleries have "nightingale" floors. The priests used to assemble in the Hall of a Thousand Tatami (straw mats). The Hojo hall is decorated with interesting paintings of the Kano period: cranes, sparrows, bamboos, flowering plums covered in snow, and snow scenes. Legend has it that the birds painted here are so lifelike that several have already flown away...

The Nishi Hongan ji

The Temple of Nishi Hongan ji (open from 10 until 4-30: visits to the apartments from 10 till 11 and 1-30 until 2-30) is still the headquarters of the Jodo Shinshu Buddhists, a sect founded in 1224 by Shinran of Higashiyama and based here since 1591. This sect grew so strong that Toyotomi Hideyashi felt it constituted a threat to the government. So he divided it into two and installed the second branch in the Higashi Hongan ji, a temple he caused to be built for this purpose.

Over the years buildings grew up around the main temple; the existing pavilions date from the beginning of the 17th century. They resemble baronial dwellings of the period; their walls are covered with delicate sculpture and paintings, masterpieces of the Kano school. The sculptures on the great gate are said to be the work of Hidari Jingoro. In the rectangular courtyard stands the Hon do, the main building, and the Daishi do. The Hon do contains a statue of Amida, one of the incarnations of the Buddha; in the Daishi do, the statue of the priest Shinran, covered with lacquer made from his own ashes, is much venerated by the faithful.

On request the monks will show the priests' quarters, situated to the south. (Unfortunately the monks speak only Japanese.) All the apartments shown are decorated with marvellous paintings: bamboos and sparrows, wild duck, chrysanthemums, horses and palm trees; on one painted door there is a picture of a cat asleep under a clump of treepaeonies. In the immense Hall of 250 Tatami audiences watched Noh plays (see: "*Theatre*"), performed on a stage in the neighbouring courtyard. In this hall Hideyoshi used to hold meetings of his Council. A painting by Tanyu depicts a Chinese Emperor receiving a Japanese Ambassador. Carved storks, the work of Hidari Jingoro, seem about to take off from the frieze above. The coffered ceiling is decorated with flowers and fans. Wooden panels depict hunting scenes, processions of little figures sheltered by immense parasols, against a background of distant hills. Beyond the halls lies a sand and rock garden and a landscape-garden.

The Higashi Hongan ji

The Higashi Hongan ji (open from 9 until 4-30) is one of the largest temples in Japan. After being burned down several times it was rebuilt in 1895 with the aid of subscriptions. The Founder's Hall has an immense wooden roof. The great wooden beams were put in place, it is said, by means of ropes woven from hair given by the faithful. Some of these ropes are visible on the south side.

The three imperial residences.

In order to visit the imperial residences permits must be requested several days in advance (passport required), from the Imperial Household Agency situated just west of the main entrance to the Imperial Palace.

The buildings that comprise the Imperial Palace (Kyoto Gosho), set in a lovely park, can normally only be viewed from outside. The first palace was built in 1788, then rebuilt in 1854 after a fire. The Seiryo den is decorated with red-lacquered pillars. The Shishin den, preceded by a series of galleries, was the scene of important ceremonies, such as coronations. The Emperor sat on his throne, behind a curtain; the court functionaries stood on the fifteen steps, in order of importance. Behind this great hall lie the imperial apartments. The Sento and Omiya palaces are used by important foreign visitors and are not on view.

The gardens of the Katsura Villa.

It was the famous master of the tea ceremony, Kobori Enshu, who created this villa and its celebrated garden, on the banks of the river Katsura, between 1620 and 1624. They were commissioned by Toyotomi Hideyoshi for Prince Hachiyo Toshito, a grandson of the Emperor. It is said that Enshu accepted the commission on three conditions: there were to be no limits as to time or money, and he was to be allowed to work undisturbed and unobserved until the job was finished. His conditions were accepted and thus the Katsura Villa was born, surrounded by the loveliest garden in the world. The whole ensemble represents "the supreme example of the architecture of a Japanese dwelling" (Yoshida) and also "a junction and a crossroads in the history of the art of the world. Here the ancient traditions of Yamato (Ise) architecture, the influence of Chinese philosophy and aesthetics of the Sung period, matured and transformed by Japanese Zen thought, and the spirit of the tea ceremony at its greatest period, all converge and transcend themselves. From here stemmed influences that have significantly modified Japanese domestic architecture and created a style which has enthusiastic disciples all over the world today. Katsura's beauty is discreet, indefinable, pure. Here there is no ostentatious luxury, nothing grandiose, nothing superficial; it is the expression of an almost heroic discipline and restraint". (Fosco Maraini)

To walk around the garden is sheer delight. No matter where one is one has the impression that the whole garden was designed to be seen from just that particular spot. The three essential ingredients, stone, trees and water—symbolising bones, hair and blood—are combined to form a marvellous living unity. Lakes, islands, delicate arched bridges with shallow steps, designed with restrictive kimonos in mind, carpets of moss, huge trees, "bonzai", stone flagged paths, lanterns —all induce a desire to stroll about, to explore, yet at the same time all are conducive to calm reflection and peaceful repose. Visitors used to leave the Shokin tei tea house, right on the water's edge, by boat. From another, the Shoka tei, set on top of a little hill, there are two quite different views: one across the lake, the other of a green and wooded landscape reminiscent of a mountain scene. A little Buddhist temple made of cypress wood has a roof of shingles, secured by bamboo pegs. Guests used to come to the Geppa ro teahouse in order to gaze at the moon reflected in the lake.

The Palace (under restoration at the moment) comprises three apartments. In the first, after crossing the halls of the Lance and the Tree, one reaches a sort of verandah, built for viewing the gardens by the light of the moon and the stars. A wall of the second is decorated with a painting: "Seven Rich Men in a Forest of Bamboo", by a member of the Kano school. The third apartment contains the enormous "Imperial Visiting Hall".

The Imperial Villa Shugaku in Rikyu

During the summer months visitors are advised to choose the early morning or late afternoon to come here, for the park has steeply sloping paths exposed to the sun, which can make for a tiring visit otherwise.

The three pavilions of this imperial villa, set on the slopes of Mount Hiei, were built during the 16th century under the Tokugawa Shogunate, on the site of the Shuga ku temple. They were designed as a retirement villa for the Emperor Go Mizuno o. The pavilions themselves are very simple and the 28 hectares of gardens blend in perfectly with the landscape. The lower and middle villas are surrounded by growing crops and ricefields. The two pavilions of the lower villa Zoroku an and Jugetsu kan, are decorated with charming paintings. The middle villa, Naka no chaya, was built as residence for the Emperor's daughter, then converted for while into a temple, known as Rikyu. It contains attractive paintings showing the carved and painted floats that are such a feature of festivals in Kyoto. A door in the villa has some pictures of carp painted on it; they are so realistically rendered that it was thought necessary to cover them with a golden net, to prevent them swimming off to the nearby lake.

It is quite a stiff climb up to the top, through the fields of crops and a succession of gates that suspicious caretakers carefully close behind the advancing visitor. However, we are now reaching the most beautiful part of the gardens. The Emperor used to come and sit and meditate here in the upper pavilion. From the balcony of "Poetic Composition" there is a view of the Dragon Lake with its two tiny islands, one, "The Island of Ten Thousand Pines" is joined to the bank by a pretty covered bridge. The whole garden is planted with carefully-pruned "bonsai". There is a path around the lake leading back to the gate to the middle garden.

Temples outside Kyoto

The Temple of 10,000 Torii, the Fushimi Inari jinja, is 3 kilometres out of Kyoto (open from sunrise to sunset). Inari is the God of Rice, one of the most popular deities in Japan. He is shown sitting on the back of a fox, traditional guardian of granaries, his messenger and companion; often he is represented by a fox alone. The Shinto shrine of Inari was founded here on this hill, during the 8th century. The "Saige Kobo Daishi" is said to have encountered here an old man laden with sheaves of rice; he recognised him to be the God of Rice

*Every day hundreds of schoolchildren
visit the country's historic sites and beauty spots,
here a party climb up to see the sunset
from the Kyumizu dera Temple, overlooking Kyoto.*

and accordingly placed the monastery under his protection.

A narrow street, lined with little souvenir shops and restaurants, leads up to the entrance to the temple. After passing under a great red torii gateway there is a flight of steps with a series of stone statues of foxes. In front the main shrine itself there is a stage, where girls perform sacred dances for the pilgrims, several times a day. Behind the shrine, leading further up the hill, there are avenues with a succession of torii gateways set very close together, each bearing an inscription with the name of the donor. These red "covered avenues" are quite remarkable and, seen at sunset, very beautiful. The Inari Temple is a very popular one and much venerated by ironworkers and cutlers, who take the God of Rice as their patron.

The Byodo in: a bird in flight

The Byodo in temple (19 km from Kyoto, open from 9 until 5) is a marvellously harmonious building, a central pavilion flanked by wings built on piles, all surrounded by trees and reflected in a small pool. It is a wooden building, painted red, with a roof of silvery tiles surmounted by a phoenix. "It evokes a bird in flight, an effect often sought by the architects of the Heian period." (Elisseeff)

During the 9th century a villa belonging to Minamoto no Tori stood on this site. It passed in due course to Fujiwara no Michinaga whose son, Fujiwara no Yorimichi, caused this splendid palace—half residence, half temple—to be built here, at the beginning of the 11th century. The central Amida Pavilion or Hoo do, Phoenix Hall, is built in the "Shinden" style, characteristic of the palace architecture of the period. Visitors are allowed, one at a time, into the central hall where the famous gilded bronze statue of Amida, the work of the sculptor Jocho (1052) is most carefully preserved. The Buddha, clad in a light robe, is seated on a lotus flower; "around him is an aureole of flying angel musicians (hiten) whose robes look like so many flames." (Elisseeff) The lower parts of the walls of the hall are covered with paintings, based on local scenes, symbolising Paradise; above there are thirty-two reliefs of delicate small musician figures.

A walk around the garden provides a succession of enchanting views of the palace, reflected in its pool. A belfrey contains a huge bell, said to be of Indian origin. This temple is depicted on the 10 yen piece.

Back to Kyoto the National Museum

The National Museum (Yamato Shichijo Kita, Higashiyama ku, open from 9 until 4-30, except Mondays) has a collection so rich that it can only be displayed one part at a time. Thus there is no guarantee that the visitor will find on view the objects he most wants to see. The museum has sections devoted to archeology, painting and sculpture. In archaeology there are objects from the Jomon and Yayoi periods on display. The Haniwa figurines (see: "Arts" and "Miyazaki") are particularly interesting. The rooms containing paintings offer a good survey of Japanese painting over the centuries; there are many "kakemonos" (painted scrolls) and a celebrated painting of Amano Hashidate by the 15th-century artist Sesshu Toyo. Finally, there are very fine sculptures of the Heian period.

The Gion and Ponto cho districts: the geishas

After six in the evening, the busy Gion and Ponto cho districts take on a special character. When the offices and temples close people crowd into the streets, and linger in the shops selling silks, fans, umbrellas, ceramics and lacquerwork. About eight the shops close in their turn and big paper Gifu lanterns of many colours are lit and hung up outside restaurants and bars. The little gardens leading to the "o chaya" or honourable tea-houses are more discreetly lit; a few sweet notes from the samisen indicate the activity within these wooden walls. At this hour the streets are not full of couples, but of men, and geishas dressed in their ravishing kimonos, with beautiful hair and make-up, trotting along at their characteristic pace.

Geishas and their life

There are sixty thousand registered geishas in Japan today, for, more than ever, their services are in demand. The visitor should not leap to conclusions here—for he is probably confusing "geisha" and "shiseid"—the latter meaning bar hostess. "Gei" means culture and "sha" person, a geisha is therefore a person practised in the arts. The profession of geisha is traditionally respected in Japan and the young ladies who follow it are highly esteemed. In former times it was only the geishas, among Japanese women, who were at all cultivated.

In order to gain admittance to an "o chaya", run by woman known as an "akami-san", it is necessary to be introduced, and the price is high. A customer never goes to a geisha's home (she may live alone or with another geisha in a "yacata"); the meeting will take place at an "o chaya" or some other fixed rendez-vous.

A geisha's apprenticeship begins at the age of seven or eight, often as a servant to an established geisha; others may sign contracts with "o chaya" or agencies and then go to special schools, attached to theatres, where they study music, singing, dancing, the tea ceremony and possibly English. Most of the pupils in such schools are the daughters of geishas. After a three year "noviciate", having reached the age of eighteen, they become "wine pourers"; as such they may serve guests with food and with sake but they are not allowed to speak to the customers. When their apprenticeship is finally over they choose their professional name ("Pacony", "First Snow"...); customers are prepared to pay a very high price for a geisha's "First Evening". The earnings of a geisha are regulated by her agency or "o chaya". First she has to pay back the cost of her education, then she is able to save. A geisha usually earns 20 to 50 dollars an hour.

Most geishas never marry, they often form a liaison with a regular client. Prominent Japanese frequently maintain a geisha as a sort of "social secretary". It is very expensive. Geishas have extensive wardrobes, luxurious kimonos and jewellery and they pride themselves on their hair and make-up. Today about a third of all geishas do in fact accept (highly lucrative) assignations outside their professional activities. Yet a geisha is never obliged, by the terms of her contract, to yield to the entreaties of a client after a meal. For such eventualities there are the "makura" ("geisha cushions") who can be summoned, if necessary, after dinner.

Any young girl who has a kimono and a shamisen (a kind of guitar) can seek a job as a "geiyu"—performing in a restaurant for 4 dollars for a three hour session. This practice grew up in Osaka; the geishas' association is highly critical of it, accusing the "geiyu" of degrading the profession of ladies of the "world of flowers and willows", the "karuykai".

A visit to Mount Hiei

The eleventh-century Emperor Shirakawa is alleged to have said: "There are three things beyond my control: the flooding of the Kamo river, success at dice and the monks of Mount Hiei."

Mount Hiei (22 km from Kyoto, access by toll-road or by train to Demachiyanagi and then by cable-railway) dominates the city of Kyoto. From the belvedere on its summit there are fine views over Kyoto and Lake Biwa. This splendid wooded

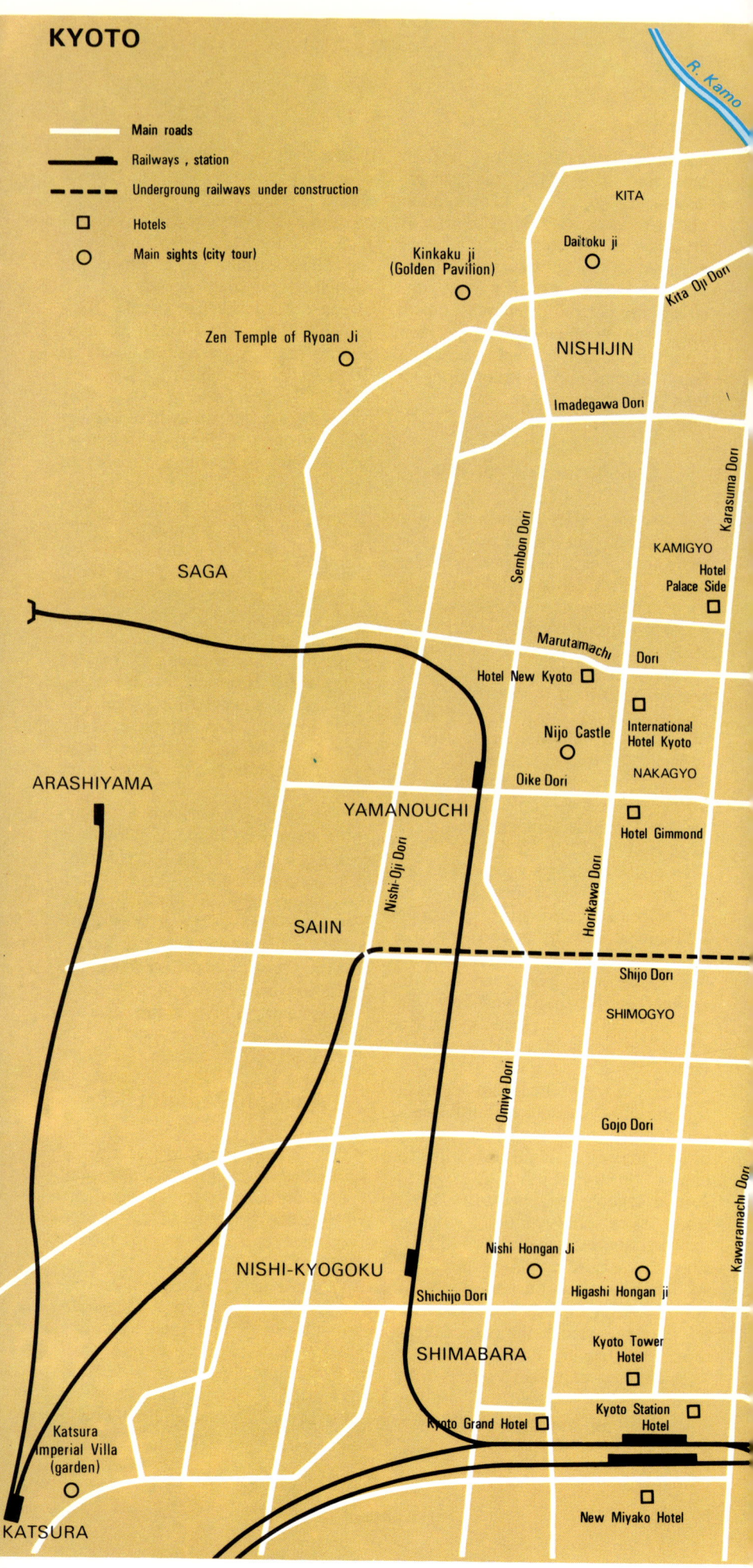

KYOTO

- Main roads
- Railways, station
- Undergroung railways under construction
- ☐ Hotels
- ○ Main sights (city tour)

R. Kamo

KITA

Daitoku ji

Kinkaku ji
(Golden Pavilion)

Kita Oji Dori

NISHIJIN

Zen Temple of Ryoan Ji

Imadegawa Dori

Karasuma Dori

SAGA

Sembon Dori

KAMIGYO

Hotel Palace Side

Marutamachi Dori

Hotel New Kyoto

Nijo Castle

International Hotel Kyoto

Oike Dori

NAKAGYO

ARASHIYAMA

YAMANOUCHI

Hotel Gimmond

Nishi-Oji Dori

Horikawa Dori

SAIIN

Shijo Dori

SHIMOGYO

Omiya Dori

Gojo Dori

Kawaramachi Dori

NISHI-KYOGOKU

Nishi Hongan Ji

Shichijo Dori

Higashi Hongan ji

SHIMABARA

Kyoto Tower Hotel

Kyoto Grand Hotel

Kyoto Station Hotel

Katsura Imperial Villa (garden)

KATSURA

New Miyako Hotel

138 KYOTO

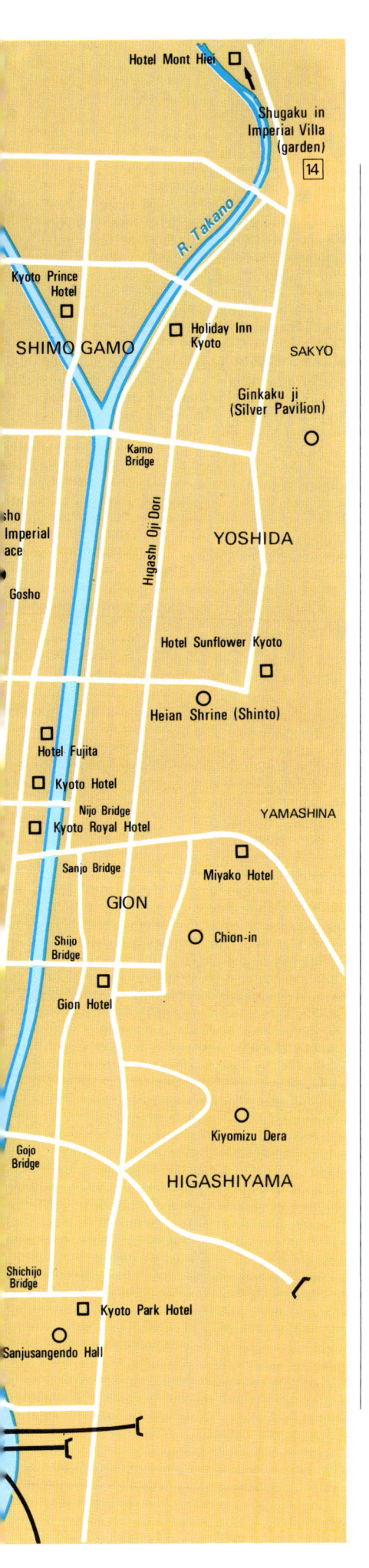

mountain—a wonderful place for walks—was, down the centuries, the eyrie from which the monks of Enryaku ji used to descend upon Kyoto. The monastery was founded in 788 by the priest Dengyo Daishi and became the headquarters of the Tendai sect.

The monastery was enormously wealthy but its pious monks soon became outnumbered by their warrior brothers. If the latter felt they had been insulted in some way bands of them would come rushing down the mountainside, armed under their habits and accompanied by their sacred palanquins lit with a thousand candles. These palanquins spread terror far and wide for it was said that any one who touched them would immediately be struck down dead. Exasperated by these troops of destructive monks, the Emperor Oda Nobunaga had the monastery pulled down; Toyotomi Hideyoshi later allowed it to be rebuilt. The present buildings are of no particular interest. They date from the 17th century and many contain statues that the public are not admitted to see. It is the view above all that makes the trip to Mount Hiei worth while.

Craftwork in Kyoto

Kyoto is celebrated for its silk and the city has more than 20,000 weavers. One can often see great lengths of silk lying out on the banks of the Kamo gawa river to dry in the sun. Craft workers are concentrated in the older parts of the city. If time is short a visit to the Craft Centre (Kumano Jinja Higashi, Sakayo ku) is a good way to see a wide selection of goods. The huge building lacks charm but its five storeys are crammed with all sorts of things for sale: there is a vast selection of kimonos, silks of all kinds and pottery. On each floor there is a small space reserved for craftsmen to actually work, so you can pause to admire an artist as he quickly paints a green bamboo branch on a sheet of incredibly thin paper, or another decorating a piece of china with a landscape scene, or a young woman making a pot, or another doing embroidery...

KYOTO

Location and access: Island of Honshu. See map and plan of the city on page 138. Pop. over 1,500,000: capital of Kyotofu Prefecture (pop. 2,500,000). Nearest airport: Osaka International Airport at Itami, 40 km—1 hr by bus or taxi. (Itami is 55 min. from Tokyo, 1 hr 45 min. from Sapporo, 1 hr 35 min. from Fukuoka). Distances: Tokyo, 513 km (2 hrs 50 min. by shinkansen—bullet train), Osaka 40 km (20 min.), Nagoya 148 km (50 min.). Post codes 600 to 606. Telephone code 075.

Travel Agency: JTB International. tel. 231-3600; behind the station.

Accommodation: Miyako Hotel****, Sanjo Keage, Higashiyama ku, Kyoto 605, tel. 771-7111, telex 5422-132, 480 rooms (28 trad. style). 20 min. by taxi from station; stands in most attractive garden. *Kyoto Grand Hotel*****, Horikawa-Shiokoji, Shimogyo ku, Kyoto 60, tel. 341-2311, telex 5422-551, 610 rooms; 2 min. from the station; fine view over the old city from 10th storey revolving restaurant. *International Hotel Kyoto****, 284, Nijo-Aburanokoji, Nakagyo ku, tel. 222-1111, telex 5422-158, 332 rooms; beautiful setting, opposite Nijo Castle, 10 min. from station by taxi. *Kyoto Royal Hotel*****, Kawaramachi Sanjo, Nakagyo-ku, Kyoto 604, tel. 223-1234, telex 5422-888, 400 rooms; 15 min. from station, by taxi, and 1 hr by taxi from Osaka Airport. *New Miyako Hotel****, 17 Nishi-Kujoincho, Minami-ku, Kyoto 605, tel. 661-7111, telex 5423-211, 715 rooms; most recently built, opposite south exit of Kyoto Station. *Hotel Mount Hiei****, Ipponsugi, Hieizan, Sakyo ku, Kyoto 606, tel. 701-2111, telex 5422-360, 75 rooms (11 trad. style); 30 min. by taxi from the station; fine panoramic views. *Hotel Fujita****, Nishizume, Nijo-Ohashi, Nakagyo-ku, Kyoto 60, tel. 222-1511, telex 5422-571, 200 rooms (18 trad. style); 10 min. from station; overlooking Kamo river. *Kyoto Station Hotel***, Higashino Toin gori, Shiokoji, Shimogy ku, tel. 361-7151, telex 5422-456, 130 rooms; opposite the station, close to JTB office. *Kyoto Tower Hotel***, Karasuma, Shichijo Sagaru, Shimogyo ku, tel. 361-3211, 150 rooms; opposite the station. *Hotel Palace Side Kyoto***, Shimodachiuri Agaru, Kamigyo ku, tel. 431-8171, telex 5423-236, 120 rooms; fine views over the gardens of the Imperial Palace. *Hotel New Kyoto***, Horikawa Maruta machi, Kamigyo ku, Kyoto 602, tel. 801-2111; 250 rooms; near Nijo Castle. *Kyoto Park Hotel***, 644-2, Sanjusangendo Mawari machi, Kyoto 605, tel. 541-6301, telex 5422-777, 60 rooms; quiet position; 5 min. from the station; close to main temples and the National Museum. *Kyoto Gion Hotel***, Horikawa Shiokoji, tel. 551-2111, 150 rooms; in the attractive Gion quarter. *Hotel Sunflower Kyoto***, 51, Higashi-Tenno cho, Sakyo ku, tel. 761-3131, 86 rooms (34 trad. style); 20 min. from the station, close to main temples. Ryokan: *Hatanaka****, Gion machi, Higashiyama ku, tel. 541-5315, 15 rooms. *Myochokaku****, Kamigamo Sakurai cho, Kita ku, tel. 791-9101, 14 rooms. *Yoshikawa Inn****, Fuya cho Sanjo Sagaru, tel. 221-5477, 12 rooms. *Inagaki****, Awataguchi Torii cho, Sakyo ku, tel. 761-1212, 18 rooms. There are dozens more ryokan of all categories.
Other hotels in Kyoto shown on the map: no. 14, *Kyoto Hotel****; no. 15, *Hotel Gimmond***; no. 17, *Holiday Inn Hotel****; no. 16, *Kyoto Prince Hotel****; no. 18, *Hotel Mount Hiei****.

Restaurants: Manyoken***, Fuyacho Higashi, Shijo dori, tel. 221-1022; grills, French cuisine. *Minokichi****, 65 Torii machi, Awatagu chi, tel. 771-4185; the oldest restaurant in Japan (250 years); sukiyaki, tempura. *Mikaku****, Kawabata dori, Shijo Agaru, Gion, tel. 561-2651; "beef restaurant", sukiyaki, Japanese style. *Izutsu****, Sanjo-sagaru, Yamato-oji dori, tel. 541-2121; grills and steaks, Japanese and Western cuisine. *Suehiro****, Kawaramachi dori, c/o Shijo dori, tel. 221-7188; established in 1900, Japanese and Western cuisine, Kobe beef, barbecue, sukiyaki. *Lipton Corner***, 377-1, Naramano machi, Nishi-iru, Shijo dori, tel. 241-3551; Western cuisine (5 branches). *Junidanya***, Hanamikoji, Gion, Higashiyama ku, tel. 561-0213; very Japanese atmosphere, Kobe beef and steaks.

Theatres: Noh theatre: Kongo Theatre, Muro machi, Shijo Agaru, Nakagyo ku. Kabuki theatre: Minamiza Shijo Ohashi Tamoto, Higashiyama ku.

Festivals: Miyako Odori, Cherry Blossom Dance, in April at Kobu Kaburenjo. Kamogawa Odori, in May and October, at the Pontocho Kaburenjo. Aoi Matsuri, 15 May, processions at the Imperial Palace and the Shimogamo and Kamigamo shrines. Mifune Matsuri, 3rd Sunday in May, Boat Festival at Arashiyama. Gion Matsuri, 16 and 17 July, parade of decorated floats. Jidai Matsuri, 22 October, parade of historical costumes of different periods.

Souvenirs: Kyoto silks.

kyushu (island of)

■ Kyushu is the most southerly of the larger Japanese islands and the third in terms of area (42,030 sq km). It is separated from Honshu by the Straits of Kammon Kaikyo which are 700 metres wide. Kyushu means "isle of fire". Four great volcanic ranges, Aso, Unzen, Kirishima, and Sakurajima, cross the island from north to south, leaving little room for the surrounding plains. Mounts Aso and Sakurajima emit sulphurous fumes...

Kyushu is also an isle of wind and water. From August to the end of September the coasts are lashed by violent typhoons often accompanied by tidal waves. There are hot springs almost everywhere which are exploited in a large number of spas.

The vegetation is lush and tropical: bamboos, exotic shrubs, and palm trees, giving way in the higher regions to cryptomerias, the most common Japanese conifer. The characteristic colour of Kyushu is a brilliant green, set off by the yellows of the bamboo, the acid green of the ricefields and the more restful green of the grassy plains. This hottest island of Japan has a heavy, humid atmosphere during the rainy season.

Kyushu has been the scene of important events in Japanese history. It was from here that Jimmu Tenno is said to have set out in 660 B.C. to conquer the other islands. Again according to history and legend the Empress Jingu, in the 3rd century, and Toyotomi Hideyoshi, in the 16th, made Kyushu their base for expeditions to Korea. During the 16th century it was the haunt of pirates who launched raids against Chinese shipping (see: *Japan through the centuries*).

Throughout Japanese history Kyushu has been the country's window on the world outside. It was from here that trade with China took place. During the 16th century the Portuguese, Spanish and Dutch all landed here; the first missionaries to Japan arrived here in Kyushu.

The island's population today is approaching 7,250,000. Fukuoka, Kitakyushu, Oita, Kagoshima and Nagasaki have become centres of heavy industry. Of all Japanese the inhabitants of Kyushu are said to be the most lively, open, emotional and welcoming to visitors.

Kuyshu receives more and more visitors every year. The Japanese come here to take cures in the spas (particularly at Beppu) which even before the war were frequented by Americans and Europeans. The coastal areas are much favoured by honeymoon couples. From Tokyo, Osaka and Kyoto there are many different organised tours—by jet, train and bus—which take in the main sights on Kyushu: Beppu, Miyazaki, Kagoshima, Mount Aso, Fukuoka and Nagasaki (see entries under these names and also the *itineraries* section in *Japanese Journey*).

MATSUSHIMA (p. 144)

Location and access: Island of Honshu (North), 10 km from Shiogama station. It is possible to make an excursion here from Sendai.

Accommodation: It is best to stay in Sendai. Matsushima has only two ryokan: *Hotel Taikanso Bekkan***, Matsushima cho, tel. 4-2161, 55 rooms. Matsushima *Dai Ichi Hotel**, Matsushima cho, tel. 4-2151, 32 rooms.

Festival: 9-11 July, Great Marine Festival.

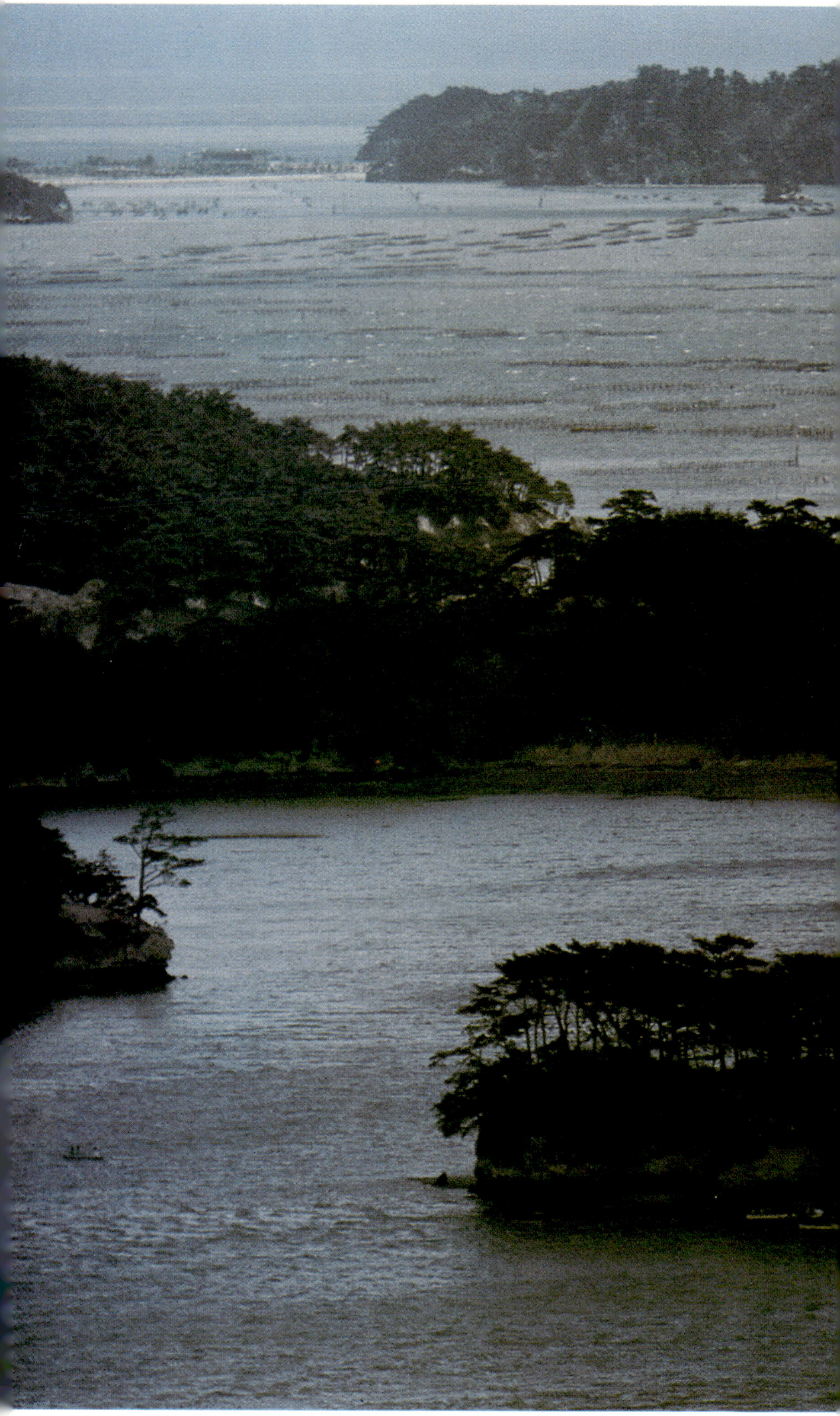

The famous site of Matsushima.
Quite close to the shore lie a host of islands and islets,
often strangely shaped and eroded by the sea,
some are crowned by temples,
others by a few twisted pines.

matsushima

■ Matsushima is one of the great national sites (sankei) of Japan, along with Itsukushima and Amano Hashidate. It consists of some eight hundred islands—varying in size from the small to the minute—scattered on the ocean quite close to the coast. Often strangely stratified, eroded by the sea, these rocky islets have taken curious forms—peaks, arches, even mushroom shapes. Some are crowned by temples, others by just a few pine trees "whose twisted trunks and branches irresistibly recall the gestures of Japanese actors". Some shelter fishing villages with brightly coloured houses, but most are uninhabited. The Japanese consider Matsushima to be most beautiful in winter, when a thick blanket of snow lends an atmosphere of unreality to its landscape. The charming little island of Oshima is joined by a little footbridge to the village of Matsushima kaigan. A longer bridge links a large wooded island to the shore. From the centre of Matsushima kaigan an avenue of cryptomerias (Japanese conifers) leads, between many curious grottoes—lived in by pilgrim monks in times gone by—to the Zuigan ji, a shrine dating from the 9th century and rebuilt in the 17th. Its pavilions, built in the "momoyama" style (see "Art in Japan") have handsome sliding doors. One hall contains the treasures; a fine bronze bell, a wooden figure of Date Masamune (a 17th-century lord), and lovely painted scrolls. Seven kilometres from Matsushima, on the little hill known as Tomi Yama, amidst a pretty garden planted with flowering cherries, stands the Shinto shrine of Daigyo.

The best way to appreciate the slightly mysterious charm of this region is to take a boat trip. There are frequent departures from Matsushima kaigan for many of the islands. The farthest of them, Kinkazan, is also known as "the Golden Flower" because of the mica in its granite rocks. *(See notes p. 141)*

SAKOURANOKI, OR CHERRY TREES IN BLOSSOM

■ *"If anyone asks who what lies at the very heart of Japan, show him the wild cherry, radiant in the morning sunlight."*
(Mozoori Norigana, 18th century)
From time immemorial, the "sakouranoki" or cherry trees of Japan have been a source of inspiration to her poets. Their immaculate blossoms symbolise joy, beauty, happy love, purity of life and delicacy of feeling. The favourite flower of the samurai, to them it stood for valour. Every year, an office of the Imperial Household organises a poetry competition with the cherry tree as its theme; thousands of Japanese take part. There are 147 varieties of "sakouranoki", purely ornamental trees, producing no edible fruit. Their flowering period—if you follow them from south to north—lasts for two months: from the end of March on the island of Kyushu to late May in Hokkaido. Some 50 million Japanese spend holidays away from home during the festival week at the beginning of May; many of them go especially to see the cherry-blossom.
In parks and gardens and temple courts you can see them, whole families who have come out for the day, to picnic, drink sake, and gaze up at the blossoms in all their transient loveliness.
"In Spring it's almost as if a sunset-tinted fleecy cloud had descended from the skies to hang upon the branches of the cherry trees." (Lafcadio Hearn)

miyajima (itsukushima)

■ Together with Amano Hashidate and Matsushima, Miyajima is one of the three great sites or "san kei" of Japan.

Its name immediately conjures up the romantic picture of a great "tori" (gateway) rising out of the water. Until the Meiji Period (1868) no one had the right to be born or to die on the sacred island of Itsuku. Women had to give birth on the neighbouring shore; the dying were hastily transported there too. It was forbidden to cut down the trees in the island's magnificent forests or to till its soil for cultivation. Today there is still no cemetary on the island; dogs are not allowed but fallow-deer roam about at will.

At the head of a little bay, backed by a wooded mountain, stands the great Shinto shrine of Itsukushima, sacred to the three daughters of Susanoo, one of whom was called Itsukushima-Hime. The temple, which dates from the 8th century, has been enlarged and restored many times over the centuries. "The various buildings, built on piles and linked together by open galleries, are so designed that the water comes right up to them at high tide. The torii is then completely surrounded by water and the whole complex of pavilions and galleries seem to float on the sea, an effect heightened by the mists, so frequent in the Inland Sea, which envelop them in a tinted veil whose colours change according to the time of day." (D. and V. Elisseeff.) At high tide the red pillars, the galleries, the thatched roofs and the carved dragons are all reflected in the waters. Low tide reveals stretches of golden sand covered with wreaths of green seaweed. The principal shrine, which consists of the main hall or "Hon den" and other halls for prayer and offerings, opens onto a wide terrace where sacred dances take place; the great torii rises beyond. Slightly to the north stands the oldest "Noh" theatre in Japan. Just behind it you can see the arches of the Sori bashi bridge.

On the other side of the river Mitarashi, stands the Treasury, which contains 12th-century Sutras (painted scrolls), a magnificent decorated chest, and a collection of "Noh" masks.

On the little hill nearby, close to a five-storey pagoda, stands the Senjo kaku or Hall of a Thousand Tatami. It was built by Hideyoshi when he was planning the invasion of Korea, in 1588. A fine wooden building with a great roof supported on massive pillars, it is said to have been built with the timber from a single giant camphor tree. A curious feature are thousands of engraved spoons which lie piled up on the tables for the offerings. In mid-July the great island festival of Kangen sai takes place, with music and a parade of illuminated boats at night. They make a fairytale picture. The avenues leading to the Shrine are lit by the hundreds of stone lanterns that line them on both sides.

Visitors are strongly recommended to spend a whole day on the island—this makes it possible to see the Shrine both at high and low tide.

MIYAJIMA (ITSUKUSHIMA)

Location and access: 22 km from Hiroshima by train; ferry crossing takes 8 minutes.

Accommodation: There are a few small ryokan on the island, but it is probably more convenient to stay in Hiroshima and then take the train (½ hr).

Restaurants: There are several small restaurants where visitors can sample an attractive variety of dishes.

Festival: The musical festival of Kangen sai takes place in mid-July.

Souvenirs: Articles made of bamboo.

*A romantically-sited torii gateway
—to one side the ocean, to the other the Miyajima Shrine,
a place of serene, almost ethereal beauty
(photo C. Salvadé).*

miyazaki

■ Visitors from Honshu feel that the South really begins at Miyazaki. The air here is warm and humid, palm trees grow along the banks of the Oyodo gawa, and the gardens are full of exotic plants. The broad avenues of the town and the seaside promenade are thronged with tourists. The great temple of Miyazaki jingu, with its blue-tiled roof, sacred to the memory of the Emperor Jimmu (660 B.C.) is much visited, as is the Heiwadai Park, with a rather ugly Peace Column (1940) and a terrace from which there is an attractive view over the city.

Scattered about the park there are amusing copies of Haniwa figures (4th-7th century). "These Haniwa figures, terra-cotta cylinders dating from the Great Tomb period, were originally designed to hold back the earth at the foot of a tumulus, thus forming an integral part of its architecture. They were decorated at the top with animals and household objects and, finally, with human figures, frequently warriors. These earthenware masks with their hollow eyes are a poignant reminder of Japan at the dawn of her history." (D. and V. Elisseeff.)

The southern coast and Aoshima Island

There are as yet no organised tours down this coast. The only way to visit it is by taxi (make arrangements via your hotel information bureau or through the JTB (Japan Travel Bureau) office.

The coast south of Miyazaki is extremely attractive, with charming creeks and inlets with curious rock formations. The little palm-covered island of Aoshima (16 km from Miyazaki by road; footbridge only), with its strange rocky platform whose ribs radiate out from the centre, looks rather like the underside of some enormous mushroom.

After passing the cactus garden of Saboten, keeping to the coast, the next place of interest is Udo jinja (43 km from Miyazaki). The red pavilions of this shrine are dispersed over the rocky cliffs, with bridges and steps hanging out over the void. It is a very beautiful spot. Many Shinto weddings take place at this shrine.

Kyushu is a favourite place for Japanese honeymooners and many couples visit Udo jinga to make an offering, leave an ex-voto, and have themselves photographed. A coin, skilfully thrown into a cleft in the rocks is said to be a good omen for the birth of a child within the year.

On the coast south of Miyazaki Province lies the Nichinan National Park.

MIYAZAKI

Location and access: Island of Kyushu. Pop. 220,000. By plane: Tokyo 1 hr 45 min. (direct), Osaka 1 hr (direct). Distances by train: Tokyo 1,487 km (1 hr via Hakata), Osaka 972 km (12 ½ hrs), Fukuoka 350 km (6 ½ hrs), Kagoshima 127 km (2 ½ hrs), Beppu 22 km (3 ½ hrs).
Post code 880 Tel. code 0985.

Information: For information about all excursions it is advisable to consult the JTB (Japan Travel Bureau), c/o Goto Shoji Building, Miyazaki eki, tel. 24-7147.

Accommodation: Phoenix Hotel***, 2-1-1 Matsuyama, tel. 23-6111, telex 7778-59, 118 rooms (incl. 22 trad. style). *Miyazaki Kanko Hotel***, 1-1-1 Matsuyama, tel. 27-1212, telex 77708, 200 rooms (97 trad.). *Plaza Miyazaki Hotel***, 1-1 Kawahara cho, tel. 27-111, telex 7779-77, 190 rooms (9 trad. style).—These three hotels overlook the river Oyodo and have mountain views; they are 5 min. by taxi from the station and 15 min. from the airport.
Seaside Hotel Phoenix***, 3083, Hamayama, Shioji, tel. 39-111, telex 7778-59, 200 rooms (96 trad. style); right on the sea. *Sun Hotel Phoenix****, 3083, Hamayama, Shioji, tel. 39-3131, telex 7778-59, 300 rooms (98 trad. style); conference room for 3,000 people with simultaneous translation available. Both these hotels are by the sea and 15 min. from the station (30 min. from the airport) by taxi.
Ryokan: *Hotel Hakuakan****, Tanigawa cho, tel. 51-1313, 80 rooms with bath; 4 min. by bus from the station. *Hotel Kandabashi****, Tachibana dori, tel. 25-5511, 110 rooms, 5 min. by bus from the station. *Tachibana Kokusai Hotel***, Tachibana dori, tel. 23-2331, 70 rooms.—These three ryokan are in the centre of the city, near the station. There are another ten ryokan of all categories.

nagasaki

■ Sited on a deep and deeply-indented harbour, surrounded by hills where houses perch high up amid flowery gardens, Japan's leading fishing port, with the biggest shipyards in the world; a modern city yet with many sleepy side-streets, temples and churches, and a Chinese quarter; everywhere pervaded by memories of the Dutch, the Portuguese, the first missionaries, not forgetting Madame Butterfly herself, of course. Such is the fascinating city of Nagasaki, the most exotic city of Japan, still deeply conscious of its remarkably varied history.

A window on the West

During the 12th century, thanks to schemes of Minamoto Yoritomo (see: *Japan through the centuries*), Nagasaki Kotaro became lord of this little harbour and gave it his name. Four hundred years later, the Portuguese, the Dutch and the Spaniards all established trading stations here. Missionaries arrived too, led by St. Francis Xavier. Their initial success in converting the Japanese led the authorities to take strong measures against them. The new master of Nagasaki, Toyotomo Hideyoshi, first of all asked the Jesuits to leave; then, judging that the Catholic converts were a source of sedition, he launched a campaign of persecution. On 5 February 1597, 26 Christians including the Spanish Ambassador were put to death by crucifixion. Persecution continued throughout the 17th century; 3,125 Christians were martyred. Yet a Christian minority clung on, in secret, in the city, until the 19th century. The Dejima district remained a refuge for Dutch traders, the only foreign trading post in the country, its only line of communication with the outside world, throughout all the time Japan remained a "closed" country.

In the mid-19th century the "opening" became somewhat wider. The study of Western culture was encouraged by the shogun; American and French ships came into the harbour, the first shipyards were opened and more traders settled in Nagasaki; the city really began flourish economically and commercially. On 9 August 1945, three days after the bombing of Hiroshima, the second atomic bomb was dropped, on Nagasaki. It destroyed a large part of the city and, according to official figures, there were 23,753 victims.

A two-day visit

Two full days at least should be spent in Nagasaki, to make an unhurried tour of its monuments and then have some time left over to stroll around the harbour and the older parts of the city and browse in the antique shops for old pottery, tortoise-shell combs or model ships. The following is a suggested programme. First day: a visit to the Belvedere in the morning, and to the old quarter and the temples in the afternoon. Second day: visit the Glover House and the Catholic church in the morning, and Dejima, the harbour, and the Peace Park in the afternoon.

Ideally one's first view of Nagasaki should be from the Belvedere on Inasa yama Hill (access by cable car or by road). From here there is a splendid view of the whole city and the harbour and out to sea. It is well worth making two trips, one by day and another by night to see the lights.

The old quarter and the temples

At the bottom of the hill, behind the tallest "torii" gate in Japan, at the top of a flight of steps, stands the Suwa jinja. The temple court is often enlivened in the morning by a "music and movement" class for tiny children; they continue their exercises unmoved by whatever is going on in the way of weddings or baptisms in the buildings around. Close by the shrine flows the Nakajima gawa, with willow trees along its banks.

The river is crossed by several stone bridges, including the famous Megane bashi or Spectacles Bridge, built in 1634 by a Chinese priest. They make it easy to walk across into the old part of the city, a delightful place in which to stroll. There are many peaceful old-fashioned streets of wooden houses, antique shops including quite a few specialising in pottery, and small craft workshops as well as tiny shops selling fans and musical instruments. Round a corner in a narrow lane stands the temple of Kofuku ji, founded by Chinese from Nankin, with its wide Hon do and delicate belfrey with an upward-curving roof. In the dark interior one can dimly discern great paintings of ships, on wooden panels, and slender fish carved in wood. Not far from here is the Kodai ji, with its statue of the Buddha, seven metres tall. Set in peaceful gardens, the pavilions of the Sofuku ji, vermilion with pretty roofs, contain an immense bronze cauldron from which poor people used to be fed in times of famine. The Confucian Shrine is an astonishingly colourful building, its brilliant roofs are decorated with birds and dragons, their bodies covered with shiny scales. Behind the shrine the steeply sloping streets are appropriately named "Dutchmen's Hill"; they lead to what used to be the foreign trading quarter. At the top of one of them stands the Furansudera (Temple of the French), the Oura Catholic Church, dedicated to the memory of the martyrs of Nagasaki.

The houses of three British traders and industrialists, Glover, Ault and Ringer, are much visited by the Japanese because of their associations with a certain (mythical) Madame Butterfly. Thanks to the progress of science it is now possible to take and escalator up the steep slope to the terrace where Cho-cho-san waited for Pinkerton's ship to return. It now boasts a statue of Madame Butterfly and a plaque to Puccini. There is an extensive view of the harbour and the vast Mitsubishi shipyards, founded in 1887, where colossal freighters and tankers are built.

THE WORLD'S LEADING FISHERMEN...

■ *"And what about the raw fish?"—this is one of the stock questions faced by travellers returning from Japan. The Japanese have acquired the reputation of being great fish-eaters; a recent American best-seller, a humorous introduction to Japan, appeared entitled "It's not all raw fish". Yet, raw or cooked, "sashimi", "sushi", crabs and seaweeds dominate Japanese cooking. Thanks to a combination of warm and cold offshore currents, the seas around Japan abound in fish. These seas furnish the Japanese with some 50 per cent of their animal protein requirements. Japan's fishing fleet leads the world, harvesting no less than 11 million tons of fish in 1976—followed by the USSR (8.6 million), China (7.3 million), Norway (3 million) and Peru (2.5 million). There have been certain difficulties in recent years, arising from pollution and disagreements with the USSR over the boundaries of fishing zones, mainly north of Hokkaido. The Japanese are actively developing all sorts of fish-farming, including shellfish and seaweeds. Their fish farms produce 10 per cent of their annual production. Special fishing grounds have been developed out at sea and along the coasts, often involving complicated works, moving rocks and earth, to provide a suitable environment for seaweeds and fish. The sea-fishing fleet consists of vessels of 10 to 100 tons. Thirty per cent of Japanese fishing is done from large vessels, outside territorial waters. There are 1,000 ton Japanese trawlers fishing as far away as the coasts of Africa and the Antarctic Ocean.*

Dejima Island

Anyone who knows Dejima from old engravings will find a visit rather disappointing. It lies about 1 ½ kilometres to the south of the station; very little now remains of the old Dutch trading station. A few warehouses have been reconstructed; there is a model of the old quarter with all its European-style houses, in the gardens down by the river Nakajima gawa, and there is a museum which has a collection of old engravings, maps, and heavy furniture and pottery of the period.

To the north of the station stands the Memorial of the Twenty-six Martyrs of Nagasaki, a bronze cross by the sculptor Funakoshi. Close by there is a little museum devoted to the history of Christianity in Nagasaki. The Peace Park, 2 ½ kilometres further north, is dominated the tall Peace Statue, commemorating the victims of the second atom bomb; it is the work of the sculptor Kitamura Seibo. If you still have hours—or days—to spend in Nagasaki the time can be most agreeably passed by strolling round the Chinese quarter and down by the harbour, absorbing the many charms of one of the most delightful cities in all Japan.

NAGASAKI

Location and access: Island of Kyushu, pop. 450,000; capital of Nagasaki Prefecture, pop. 1,588,000. Distances by train: Fukuoka 154 km (2 ½ hrs), Osaka 778 km (7 hrs), Tokyo 1,331 km (10 hrs). Airport at Omura, 40 km from the city: Osaka 1 hr 15 min., Tokyo 1 hr 50 min. Post code 850. Tel. code 0950.

Travel agency: JTB, Nagasaki eki, tel. 23-1261 and JTB, opposite the station, tel. 26-4922.

Accommodation: Nagasaki Tokyu***, 18-1 Minamiyamate machi, tel. 25-1501, 26 rooms. *New Nagasaki****, 14-5 Daikoku machi, tel. 26-6161, 60 rooms. *Park Side****, 14-1 Heiwa machi, tel. 45-3191, 67 rooms + 5 trad. style. *Nagasaki Grand***, 5-3 Manzai machi, tel. 23-1234, 123 rooms + 3 trad. style. *Nagasaki Heights***, 3-19 Kozen machi, tel. 22-3156, 22 rooms + 28 trad. style. All the above hotels are in the city centre, 5 min. from the station by taxi, and 1 ½ hrs from the airport.

Ryokan: *Hakuunso****, Kajiya machi, tel. 26-6307, 40 rooms. *Nagasaki Kanko Hotel Shumeikan****, Chikugo machi, tel. 22-5121, 65 rooms. *New Hotel Chuoso****, Manzai machi, tel. 22-2218, 26 rooms. *Yataro****, Irabayashimachi, tel. 22-8166, 58 rooms. *Nagasaki Kokusai Hotel Nisshokan***, Nishikazaka machi, tel. 24-2151, 72 rooms. As well as some twenty two and one star ryokan.

Youth Hostels: Nagasaki, Tateyama cho; 90 beds. Nagasaki Oranda zaka, 2 Oura machi, tel. 22-2730, 55 beds.

Restaurants: Grill Moon** (Chinese), 6-2 Shirohae cho, Sasebo, tel. 3-511. *Chisan** (Western cuisine), Manzai machi, tel. 26-9277. *Shikairo** (Chinese), Kago machi, tel. 22-1296. As well as dozens of other Japanese and Chinese restaurants.

Festivals: Suwa Shrine Festival, 7 and 9 October.

Shopping and souvenirs: In the Higashi Hamano dori and Hamaichi dori (streets) and the Hama ma chi (covered arcade) and in the hotel shopping arcades. Tortoise-shell objects, coral jewellery, cultured pearls, antiques.

*A Mecca for admirers of Madame Butterfly,
Nagasaki rises high above its sheltered roadstead.
The secrets of its fascinating history
await the visitor amid the streets
of this most Oriental of all Japanese cities
(photo C. Salvadé).*

nagoya

■ Nagoya, also known as Chukyo, "Capital of the Centre", is the third largest city of Japan and has a population of about three million. It is not a major tourist sight and need not detain a visitor for more than a few hours.

The city developed late. Tokugawa Ieyasu (see: *Japan through the centuries*) built a castle here that was lived in by his family until the Meiji Restoration in 1868. It was about then that the city really began to develop economically and industrially. These developments were facilitated by its site, on the broad Nobi plain, and its situation on an axis between Tokyo, Osaka and Kyoto; the building of the railway was very important as well. Nagoya harbour is artificial, it had to be deepened and then protected from typhoons by breakwaters, but it has gradually become the third most important port in Japan, after Kobe and Yokahama.

Nagoya Castle was destroyed by bombing during the Second World War but was rebuilt and opened again in 1959. Owned by the Tokugawa family until 1867 it became first an Imperial residence and then the property of the city. Its proud five-storeyed keep, with golden dolphins on the top, dominates Nagoya. It contains a museum with an interesting collection of weapons and some fine painted sliding doors. The ideal time to visit the castle gardens is the autumn, when the chrysanthemums are in flower.

The Atsuta Shrine, said to have been founded in the 1st or 2nd century, is the repository of the sword, one of the three emblems of the emperors of Japan (the others, the mirror and the jewels, being at Ise and in Tokyo). This sword was the gift of the goddess Amaterasu to Yamato Takeru (see "*Religion*"). Atsuta is one of the oldest and sacred shrines in Japan. In its great wooded park there is a series of tea-houses, set about a little lake. Pilgrims are not allowed to cross the Hai den, close to the sacred enclosure, Nakanoe, in the centre of which the main building, the Hon gu. The shrine's treasure (jewels, masks, weapons, paintings) are kept in the Bunka den.

The Tokugawa Art Museum (open from 10 till 4, except on Mondays) contains objects that belonged to the Tokugawa family as well as thirteen scenes from a painted scroll or "emakimono" of the famous "Tale of Genji".

In a pavilion near the Nittai ji temple are five hundred painted figures of the deity Rakan; they date from the early eighteenth century and are remarkable for their diversity and liveliness.

NAGOYA

Location and access: Island of Honshu, pop. 3,000,000; capital of Aichi Prefecture, pop. 5,950,000. Airport at Komaki, 13 km from the city; flights by ANA to all major Japanese cities and JAL flights to Fukuoka and Tokyo 3 times a week. Distances: Tokyo 342 km (50 min. by air, 2 hrs by train), Osaka 75 km (1 hr by train), Gifu 39 km (½ hr by train). Post codes 450-466. Tel. code 052.

Travel agency: JTB, on the station square, to the left as you leave, opposite JAL, tel. 563-2076.

Accommodation: Nagoya Miyako Hotel****, 2 Nishiyanagi cho, Nakamura ku, tel. 571-3211, 400 rooms; opposite the station. *Nagoya Terminal Hotel****,* 18-1 Sasashima cho, Nakamura ku, tel. 561-3751, telex 445-7263, 190 rooms; on the station square; 30 min. by taxi from the airport. *International Hotel Nagoya***,* 3-23-3 Nishiki, Naka ku, tel. 961-3111, telex 444-3720, 260 rooms. *Meitetsu Grand Hotel***,* 1-223 Sasashima cho, Nakamura ku, tel. 582-2211, telex 442-2031, 240 rooms; 5 min. from the station and 20 min. from the airport, by taxi. *Hotel Nagoya Castle***,* 1-15 Hinokuchi cho, Nishi ku, tel. 521-2121, telex 445-2988, 250 rooms; the most pleasantly situated, right opposite the Castle. *Nagoya Kanko Hotel**,* 1-19-30 Nishiki, Naka ku, tel. 231-7711, 505 rooms; opposite the station. *Hotel New Nagoya**,* 4-1 Horiuchi cho, Nakamura ku, tel. 551-5131, telex 442-2061, 86 rooms; opposite the station.

nara

■ The city of Nara lies in the centre of the Yamato basin. In the seventh century Nara was the first proud capital of Japan; today it is a quiet university city, famous for its craftsmen (calligraphers and wood- and lacquer-workers), an agricultural centre and proud possessor of a fine Deer Park in which stand some of the country's most ancient monuments.

In 710 the Empress-Regnant Gemmyo broke with the Shinto tradition that dictated that the capital moved (for reasons of ritual purity) on the death of a ruler and established a permanent capital at Nara. Strongly influenced by the enthusiastic accounts of travellers to China and Korea the new city was modelled on the T'ang capital of Tch'ang-an (present-day Sian) and under seven successive sovereigns was the influential centre of Chinese culture in Japan. Under aristocratic patronage Buddhism, first introduced into Japan in the sixth century, flourished during the seventh and eighth; Buddhist culture reached its peak here at Nara.

Temples and palaces were built by Chinese architects, painters and sculptors flourished and the whole court studied Chinese language and calligraphy. Monasteries became increasingly important and were actively involved in court intrigues. A monk named Dokyo became the advisor and the lover of the Empress Koken but was exiled when it became evident that he had designs on the throne. The Emperor Kammu (736-805) felt that clerical influence was becoming excessive and decided to move the capital to Kyoto, then known as Heian.

The main temples are to be found in a vast park (528 hectares). The Temple of Kofuku-ji, headquarters of the Hosso Sect, stands close to the Sarusawa Pool in whose waters its five storey pagoda is reflected. It was built in the eighth century by the Fujiwara family and became extremely prosperous, rivalling the famous monastery of Enryaku-ji on Mount Hiei at Kyoto. To the north of the pagoda stands the main building, To Kon do and the Shu Kon do which houses a very beautiful seventh-century bronze head of great serenity and purity of line. To the north-east of the To Kon do is the Kokuho-kan Museum which contains part of the art treasures of the Kofuku-ji, notably sculptures of the Unkei school (see: "*Arts*").

The largest wooden structure in the world

The Todai-ji (East Great Temple) was founded in 743 by the Emperor Shomu. Its principal building, the Daibutsu-den, built to house a great statue of the Buddha, is said to be the largest wooden structure in the world. The monumental entrance, 29 metres high, with one floor above it, the Nandaimon, houses two statues of the Deva kings, guardians of the temple, dating from 1203, the work of the sculptors Unkei and Kaikei. "Half-naked and holding a "vajra" (diamond, thunderbolt, weapon) one figure has his mouth closed, to symbolise contained power, the other has his mouth open, to symbolise power in action." On the north face of the doorway are two carved lions.

An avenue leads to the Chu-mon, linked by a corridor to the Daibutsu-den (Great Buddha Hall) a vast wooden building 57 metres long and 48 metres high which stands reflected in a pool. The gilded ornaments on the ends of the roof are called "shibi"—the tail of the "shi" bird and are reputed to protect the building from fire. The Daibutsu was finished in 751 and ceremonially opened in the presence of the Emperor and 10,000 monks.

The building has suffered over the centuries from earthquakes and fires. Restoration is under way at the moment. In front of the temple stands a beautiful octagonal bronze lantern. The bronze statue of the Great Buddha (22 metres tall and weighing 751 tons) dates from the eighth century and took five years to cast. It was re-cast in the sixteenth century and is taller than the Great Buddha of Kamakura but not so fine.

To the west of the Daibutsu-den stands the Kaidan-in, the platform where priests were ordained. It is the oldest building of this kind in Japan. The statues of the four Devas or celestial guardians are made of dry lacquer with obsidian eyes. North of the Daibutsu-den is the Shoso-in, an eighth-century wooden building which houses the collections of the Emperor Shomu, works of art from India, China and Korea, as well as the oldest example of Japanese painting of the Nara period, the screen of the Beauties under the trees. To the east, the Sangatsu-do houses some fine sculptures including one of Fukukjaku Kannon in dry lacquer and wearing a crown containing 20,000 pearls.

In the middle of the park is the Nara National Museum (open from 9-30 until 4, closed on Mondays) containing works from various temples in the city. It contains fine sculptures in wood and bronze. Among its paintings are some "makimono" (painted scrolls) including the "makimono of Honen Shonin Eden" which dates from the early fourteenth century; its forty-eight scrolls depict the life of the painter Honen.

The Kasuga shrine: 3,000 lanterns

On the slopes of the Wakakusa yama hill, in an extensive deer park, stands the Kasuga taisha, the most important Shinto shrine in the region. Founded in the eighth century, the vermilion and white buildings of the sanctuary are surrounded by some 3,000 lanterns of stone, iron and bronze; the oldest of them date from the eighth century and are all lit on the night of 3 and 4 February and 15 August—a most attractive sight.

Passing through two gates and along an avenue lined with lanterns we reach a courtyard where there are four shrines and a remarkable tree, a graft of maple, cherry, camelia and wisteria. In the Treasure House there is a collection of weapons, armour and masks for ritual shrine dances.

The Buddha of the Yakushi-ji temple

The temple of Yakushi-ji, 5 kilometres from Nara, was built in the seventh century by the Emperor Temmu for the recovery of his sick Empress. Its main hall contains the wooden image of Yakushi Nyorai, the Buddha of medecine. Nearby stands the To-to Pagoda (seventh century) with its three double stories; it is reckoned to be the oldest example of pre-Nara period architecture. It is crowned by figures of flying angel musicians. (This temple should not be confused with the Shin Yakush-ji, a small temple near the Kasuga shrine.)

The halo of a thousand Buddhas

During the eighth century the Chinese priest Ganjin founded the Toshodai-ji temple (4 km from Nara); an elegant building influenced by Chinese architecture of the T'ang period. In the main hall, surrounded by colonnades stands the gilded dry lacquer figure of Birushana Butsu, with a halo decorated with a thousand tiny Buddhas. The Miei building houses an eighth-century dry lacquer figure of Ganjin, the work of one of his followers. The master is seated, his eyes half-closed, in an attitude of calm meditation—though as some would have it the artist has depicted Ganjin as blind. This work is the oldest known lacquer figure. The paintings on the sliding doors, "fusuma" are the work of the famous contemporary artist Kai i Higashiyama (see: "*Arts*").

Horyuji, a masterpiece of Buddhist art

At Ikaruga, 12 kilometres from Nara, in a peaceful setting of trees, stand the grey tiled buildings of Horyu-ji, an elegant pagoda and the famous "dream pavilion". Horyu-ji,

*The peaceful city of Nara,
cradle of Japanese Buddhism and
the first centre of Chinese influence,
a proud capital during the 8th century,
now a centre of crafts and agriculture.
Here are some of the country's most venerable monuments.*

NARA 155

the oldest temple complex in Japan, contains the oldest wooden buildings in the world. Founded at the end of the sixth century by the Buddhist Prince Shokutu, the temple is said to have been burnt down in 670 and re-built between 679 and 693. The complex consists of forty buildings, the Eastern and Western Monasteries—the latter being the more important.

A wide avenue, lined with pine trees, leads the visitor to the southern gate and then to the middle, double-storied gate, the "Chu-mon" where there are two carved wooden guardians in niches—one painted red (to symbolise light) and the other black (to symbolise darkness). The figures date from 711 and are the oldest of this type in existence. From the Chu-mon two galleries lead towards the Dai-ko-do building, forming a courtyard in which stands the five-storied Kon-do pagoda.

This is an elegant structure, built on a stone foundation. The four walls of its bottom floor are covered with scenes from the life of the Buddha in terracotta. The main hall, the Kondo, stands on a stone base. Its roof projects out from the supporting columns on eighteenth-century props in the shape of climbing dragons. Unfortunately the splendid frescoes on the walls, depicting the "paradise of Amida" were badly damaged by fire in 1949. The Kondo contains the famous sculpture of the "Gakya trinity" which dates from 623; a gilded bronze, it depicts the Buddha, seated, with a great leafy halo.

To the north of this compound is the Great Lecture Hall where there are fine statues. To the east is the modern Treasure House where marvellous works of art from the temples of Horyuji are displayed. The bronze statue of Kannon, goddess of mercy, (eighth century) was said to transform nightmares into sweet dreams. The image of the Kudara Kannon, carved in camphorwood is much admired for its slender elegance. The Tamamushi no zushi, a reliquary decorated with gilded scarabs, belonged to the Empress Suiko (554-628). Another reliquary bears an Amida trinity in bronze, supported on lotus blossoms, evoking the flowers that grow in the sacred pools. One such pool is depicted on a bronze plaque, engraved with waves and lotus leaves.

In the eastern shrine stands the Yumedono or Hall of Dreams, the oldest octagonal sacred building in Japan. It was here, according to legend, that Prince Shotoku used to receive in dreams the answers to his problems. This building, considered one of the loveliest in the country, contains the statue of Yumedono Kannon.

To the north-east of the Horyuji complex lies the temple of Chugu-ji and a convent with a statue of a Bodhisattva meditating in the lotus position; this Miroku-Bosatsu radiates a marvellous aura of peace.

NARA

Location and access: Island of Honshu. Pop. 200,000. Capital of Nara Prefecture (pop. 570,000). Kyoto 44 km, Osaka 33 km, Tokyo 490 km. JNR station and private rail links with Kyoto and Osaka. Coach tours from Kyoto and Osaka.
Post code 630. Telephone code 0742.

Travel agency: JTB, Kitagawa Building, Nishi Gomon cho.

*Accommodation: Nara Hotel*** 1096 Takabatake, Nara city 630, tel. 26-3300, 75 rooms (140 beds); wooden building, Japanese palace style, situated by the lakeside in the park. *Hotel Yamatosanso*** 27 Kawakami cho, tel. 26-1011. 9 rooms (+ 42 Japanese style rooms); close to temple.
Ryokan: *Nara Kokusai Hotel**** Omiay cho, tel. 23-6001. 34 rooms. *Nara Park*** Horai cho, tel. 44-5255. 27 rooms. *Kasuga Hotel*** Noborio-ji cho, tel. 22-4031. 57 rooms.

Festivals: 3-4 February, Kasuga Shrine, Lantern Festival. 2 May, Tode-ji, Emperor Shomu Festival. 12 May, Nigatsu do, "Omizutori" Festival. 15 August, Kasuga Shrine "Bon Matsuri" Festival.

Souvenirs: Lacquerwork.

naruto

■ Off eastern Shikoku, a narrow strait 1,300 metres across, joins the Pacific Ocean and the Inland Sea. It lies between the island of Oge (linked to Naruto by a suspension bridge) and the island of Awaji Shima. A difference in level of about 1 ½ metres combined with the narrowness of the gap gives rise to remarkable whirlpools, almost 15 metres across, at each high tide—particularly in spring and autumn. On the top of the hill on Oge there is a belvedere which gives an excellent view of these seething waters, a very impressive spectacle when the tide is really high. You can get a closer view of the whirlpools from the motor boats that ply between Naruto and Awaji shima.

NARUTO

Location and access: Island of Shikoku (north). Distances: Tokyo 792 km (by train, ferry at Uno), Matsuyama 215 km, Takamatsu 73 km. Buses run to Takamatsu (2 hrs), leaving from in front of the Grand Hotel. Telephone code 08868.

Accommodation: Ryokan only. *Mizuno***, Naruto cho, tel. 6-4131, 33 rooms (17 with bath). *Naruto Koen Hotel***, Naruto cho, tel. 7-0211, 27 rooms (4 with bath). *Schichishuen**, Naruto cho, tel. 6-5161, 19 rooms (2 with bath); this simple inn has a dramatic view of the whirlpools.

Festival: Awa-Odori, in mid-August.

Souvenirs: Attractive bamboo boats.

NIKKO

Location and access: Very popular National Park in Central Honshu. Access from Tokyo: by JNR train, from Ueno Station (145 km, 2 hr 10 min.), or by Tobu Railway, Nikko line, from Asakusa Station (140 km, 1 hr 45 min.). By car, take the Nikko Motorway, also used by the tour buses from Tokyo.

Accommodation: Nikko Kanaya Hotel***, Kami-Hatsuishi cho, Nikko, Tochigi Prefecture 321, tel. (0288) 4-0001; beautifully situated in the centre of the park, close to the temples. *Nikko Lakeside Hotel***, 2482, Chugushi, Tochigi Prefecture 321-16, tel. (0288) 5-0322. 30 min. by taxi from the station; beautifully situated in the centre of the park, close to Lake Chuzenji.

Festivals: Spring Festival, procession of a thousand samurai, 18 May. Autumn Festival, big procession.

nikko

■ "Nikko wo minai uchi va, kekko to iu na", says the Japanese proverb,—"He who has not seen Nikko has no right to use the word magnificent".

"Here, on the slopes of holy Mount Nikko, in the middle of a dense forest, with the unceasing sound of waterfalls coming through the cedars, there is a succession of enchanted temples of bronze, of lacquer and of gold which seem to have been wafted here by some magic wand and taken root amid the ferns and moss, under the great branches, in this wild and lovely place. This sudden great splash of gold in the depths of the mysterious forest makes these tombs unique. This is the Mecca of Japan..." (Pierre Loti, "Japoneries d'automne").

"Nikko is exhaustion, old age, limpness, a false kind of beauty desperately trying to impress with gold, grandeur and astonishing and excessive effects; Nikko has the smell of death about it." (F. Maraini)

A wedding cake

Confronted by such contrasting views the visitor may well wonder whether to make the journey to Nikko. He should not hesitate, for it is an extraordinary experience. Opinions may differ about "this elaborate set piece, this wedding cake" (Maraini), but it is impossible to be unmoved by the beauty of its setting, in which the splendid trees (cedars, birches, maples and cherries), streams and rocks all harmonise so perfectly. One often thinks of Japan as a place of refined, almost austere beauty—but there is another aspect too, reflecting a desire for luxury, abundance and a riot of colour; Nikko has both.

Apart from the town the name Nikko embraces the whole mountain region that surrounds the Shoguns' tombs. The first small Buddhist temple was built here in the 8th century, by the priest Shodo Shonin. The Rinno ji, called first Mangan ji, was rebuilt during the 17th century here at Futara yama, or Nikko in Chinese. From the late 9th century on, Emperors caused a succession of temples to be built here. But Nikko's real period of glory began in 1616, when the second Tokugawa Shogun began to build the mausoleum of his father, Iyeyasu, who had expressed a wish to be buried here. Iemitsu, himself later to be buried at Nikko, began the Tosho gu in 1634. At the Meiji Restoration the shrine narrowly escaped destruction as it had had such close associations with the Shogunate. However, it was spared and was declared a National Treasure.

An avenue of cedars, thirty kilometres long, used to link Utsunomiya with Nikko. The river Daiya is crossed by the elegant "Sacred Bridge", covered with red lacquer heightened with gilding. Built by the Shogun during the 17th century this bridge was not formerly open to ordinary mortals; it is now open during festivals.

After the bridge, on the right, stand the red Hon gu and the three-storeyed pagoda known as Shihonryu ji. Left of the bridge is a broad paved path that leads, beneath a vault of cedars, to the Rinno ji (formerly Mangan ji). If permission is asked it is possible to visit the priests' quarters in the Hon bo, and the gardens, as well as to see a very fine gold-lacquered altar. The Sambutsu do, built in 1648 at Mount Hiei (Kyoto) and brought here later, contains three large statues of the Buddha. The Staircase of a Thousand Men, where the common people used to stand during festivals, is dominated by a high torii (gateway) made of granite, the offering of a 17th-century daimyo (lord), and by a graceful five-storeyed pagoda, decorated with the signs of the zodiac and with handsome black lacquer doors.

The Tosho gu, the most important shrine, it took thirteen years to build, is entered through the Omote mon Gate which bears all kinds of motifs —lions, tigers, elephant heads, as well as flowers, apricots—and wild ducks. The courtyard inside is enclosed by a wooden wall painted bright red. There are various small buildings used for ceremonies. One of them bears two carved and painted elephants, said to be the work of the sculptor Hidari Jingoro, from drawings by Tanyu. Their hind legs are strangely twisted. A door nearby bears carvings of the Three Wise Monkeys—"See no evil, hear no evil, speak no evil".

Three monks: Three "sages"

The Shrine Library, across the courtyard, contains 7,000 volumes. Steps lead up to a second courtyard, surrounded by a stone balustrade, where there is a grand bronze lantern and a bell, both from Korea, a huge candelabrum, a gift from the Dutch during the 17th century, as well as a hundred lanterns, the gifts of various daimyo. On a terrace above stands the Yakushi temple, rebuilt after a fire in 1961; its interior is richly decorated with gold and coloured lacquer. The monumental gateway, Yomei mon, is a masterpiece of luxuriating richness; it leads to the third courtyard. The carved columns are painted white. In the medallions are two tigers, in which the grain of the wood is skilfully used to render the fur. On one of the columns the designs are deliberately painted in reverse to evoke the jealousy of the gods at such a feat of perfection. There are carvings of birds, Chinese sages, black-painted dragons, and children playing, in the medallions of the doors themselves and in the galleries.

They are said to look particularly fine in the light of the setting sun. Only high-ranking samurai used to be allowed to pass through the Yomei mon.

The next courtyard is surrounded by the Mikoshi gura, where the sacred palanquins are kept, they are used during festivals.

There is a stage here for sacred dances. The walls of this courtyard are capped with bronze.

The Kara mon, Chinese Gate, is supported by pillars of rare woods,

noboribetsu

brought from China, carved with dragons, bamboos and plum trees. The Hai den shrine consists of three halls. The doors have portraits of the "Thirty Six Poets" on their lintels. At the end of the main hall is the "Sacred Mirror". A line of "tatami" leads to the Hon den, which also consists of three halls. The last of these contains the gold-lacquered shrine sacred to the spirits of Iyeyasu, Hideyoshi and Yoritomo. Iyeyasu's body does not lie in this sumptuous temple itself. To find his tomb you have to retrace your steps through the Yomei mon and the Kara mon, pass through a gateway decorated with the "Sleeping Cat", a famous sculpture by Hidari Jingoro, and then climb two hundred steps. The tomb itself is a little bronze and gold pagoda. From the five-storeyed pagoda an avenue leads to the Futurasan. From here steps lead up to the temple and tomb of Tokugawa Iyemitsu.

The visitor to Nikko should not limit himself to seeing the famous temples. A few hours at least should be spent in the splendid park crossed by the road to Numata (90 km). After passing the Nikko Botanical Gardens (an interesting collection of plants and animals) and the waterfall, there is a steep climb with no less than forty-eight hairpin bends (a one-way road, mercifully; there is another for the return journey on the other side of the Kegon Gorge). There are many look-out points with magnificent views, before arriving at Lake Chuzenji ko, a popular spot for boating. At the far end of the lake are the 200 metre Kegon Falls. The whole region is particularly lovely in spring, when the azaleas are in flower, and in the autumn, when the maples are turning colour. From the village it is possible to climb the mountain of Nanzai san. Many pilgrims make the ascent, which used to be forbidden to women. Other glories of Nikko National Park include the Dragon's Head Falls and the wild western slopes of Nanzai san.

(See notes p. 157)

■ Noboribetsu is no longer the charming, picturesque little place that Mousset portrays in his novel.

It lies a few kilometres from the industrial belt, on the southern coast of Hokkaido, in the valley of the Noboribetsu gawa. The most important spa in Hokkaido, Noboribetsu consists of a long, winding street, lined with hotels and traditional souvenir shops all selling the traditional local souvenirs—carved wooden bears of all sizes. As early as the fifteenth century the iron and sulphur springs here were known for their curative properties. Every days coaches disgorge hordes of tourists and schoolchildren who have come here to see the Jigoku dani or Hell. This is a bow-shaped valley, with brightly-coloured earth and bounded by reddish peaks, full of steaming sulphur springs. Noboribetsu is a centre for walks and excursions. There is a cable car from the village up to the top of Mount Shihorei where you can take a coach to the lovely lake of Kuttara ko, 4 kilometres further on.

You can take a tour, or simply a train, from Noboribetsu to the village of Shiraoi (30 km) where there are Ainu people, living in their traditional huts, who receive hundred of Japanese visitors every day. The arrival of every coach party is greeted by them with dances and songs. A magnificent bearded old Ainu sits in his hut and talks to curious visitors through a microphone! The approach to the village is through a forest of souvenir shops offering all sorts of mementos, mostly made of wood. There is a little museum with displays showing how the Ainu people used to hunt and fish.

NOBORIBETSU

Location and access: Island of Hokkaido. Distances by train: 207 km from Hakodate, 91 km from Sapporo.

*Accommodation: Noboribetsu Grand Hotel****, 154 Noboribetsu Onsen, Hokkaido 059-05, tel. (01438) 4-2101, 100 rooms + 140 trad. style. Close to the Jigokudani springs.

Souvenirs: Wooden articles.

okayama

■ Okayama is a large industrial city with an important harbour; it is dominated by its vast black castle, known as the "Castle of the Crow". Built in the 16th century, the castle was owned in turn by the Ukita, Kabayaka and Ikeda families; it was destroyed in 1945 and rebuilt in 1966. The principal sight near Okayama are the gardens of Koraku-en, among the most famous in Japan; they are situated on an island in the middle of the Asahi gawa. Begun in 1687, finished in 1700, they were the property of the Ikeda family until 1887 when they were left to the city. In 1922 they were declared to be a National Treasure.

The gardens were first called "Chayayashiki" (Garden of the Tea Bushes) because tea grew so abundantly there, then Koen (Behind the City) and, finally, Koraku-en (Place of Pleasures behind the City). They were laid out according to the rules established by the great landscape architect, Enshu, during the Edo period. A feature of the gardens are the perspective effects obtained by the skilful use of two small hills. A lake with tiny islands symbolises the Inland Sea. This lake, surrounded by bamboo woods, serves as a focus for all the essential elements of a Japanese garden: waterfalls, streams, little bridges—some built in zig-zag fashion, fountains, shrubs and bonzai pines. There are also tea-houses and little temples.

OKAYAMA

Location and access: Island of Honshu (central). Pop. 515,000; capital of Okayama Prefecture, pop. 1,900,000. Distances by train: Kobe 143 km, Hiroshima 163 km, Tokyo 708 km. Okayama Airport is 8 km from the city; Tokyo is 2 hrs by air. Post code 700. Telephone code 0862.

Accommodation: Okayama Kokusai Hotel****, Kadota, tel. 73-7311, telex 5922-669. 200 rooms; very pleasantly situated, in Higashiyama Park; 10 min. from the station and 15 min. from the airport, by taxi. *Okayama Plaza Hotel****, 116 Hama, tel. 72-1201, 85 rooms; very pleasantly situated, near Koraku en Gardens and Okayama Castle; 10 min. from the station and 20 min. from the airport, by taxi. *Hotel New Okayama****, opposite the station, tel. 23-8211, 80 rooms. *Okayama Royal Hotel****, 2 Ezu *(See other notes p. 162)*

osaka

■ Almost completely destroyed during the Second World War, Osaka was rebuilt according to the latest ideas of modern town-planning. Its tall buildings and business quarters are encircled by ring-roads built on three levels, so you have motorways that are sometimes fifteen or twenty metres above ground level. Osaka's residential areas are rather dull, criss-crossed by wide avenues at regular intervals. The city offers a view of a rather dull aspect of Japan that the visitor would perhaps prefer to leave unexplored, yet its importance is undeniable in the country's general economy, and even this rather dreary city contains buildings that are well worth seeing.

From Naniwa to Osaka

According to legend the Emperor Jimmu landed in Japan, in 660 B.C., near the village of Naniwa ("swiftly-flowing wave") at the mouth of the Yodo gawa. For three centuries the Emperors lived at Naniwa and improved the locality by building drains and canals. From this time on commercial links were developed with Korea.

In the 16th century Naniwa became Osaka ("great coast") and gained considerable importance under Toyotomi Hideyoshi who had a castle built here. Tokugawa Iyeyasu successfully laid siege to the city in 1615.

During the Edo period Osaka was a great commercial centre. The daimyo came and sold their rice here (their samurai paid feudal dues in rice), to the merchant gilds. The merchants sold it in turn in other cities, principally Edo (Tokyo). The merchants of Osaka grew prosperous and became important patrons of the arts.

At the end of the 19th century the opening of the port to foreign trade greatly assisted the economic expansion of the city.

Today Osaka is Japan's third most important port.

A day in Osaka

The imposing Castle symbolises the arrogance of Toyotomi Hideyoshi, who built it on the site of a former temple, the Hongan ji. More than 100,000 workers toiled to build it. Under the Tokugawa Shogunate (see: *Japan through the centuries*), the castle was lived in by a Governor; it was partly demolished after the Meiji Restoration (1868). The existing donjon was rebuilt after the war. The gardens and the donjon (42 metres) are enclosed by ramparts made of great blocks of stone, the largest is said to weigh 520 tons! The "Cherry-tree Gate" leads into the inner courtyard and the donjon. Behind the castle is the Municipal Museum which contains an interesting collection relating to the history of Osaka.

After looking at the Castle we suggest that the visitor takes a taxi to the Shitenno ji Temple. Founded by Prince Shotoku at the end of the 6th century, this temple used to be one of the oldest in Japan; it was totally rebuilt after the war. In the great courtyard stands a five-storeyed pagoda, the main hall or Hondo and the preaching hall. The Treasury contains a statuette of the Goddess Kannon, made of gilded brass.

In the centre of Tenno ji Park is the Municipal Art Museum (open from 9 until 5) which contains a fine 13th-century painted scroll illustrating the "Tale of Genji" (see: *Literature*).

There is probably time left for a quick taxi ride to see the Sumiyoshi Shrine. It was founded, according to legend, by the Empress Jingu after her return from Korea (3rd century), as a thank offering to the gods for her safe return during a storm. The existing building dates from 1800. It is decorated with numerous stone lanterns, gifts from grateful sailors.

An appropriate way to end a day in Osaka might be to do some shopping in the underground Shinsaibashi suji Arcade; there is a wonderful selection of things to buy. In the evening there is always the possibility of seeing a "Noh" play or attending the Kabuki or Bunraku Theatres (the latter being an Osakan creation).

If more than one day is to be spent in Osaka then the Fujita Museum (Ajima cho, Miyakojima ku; open from 10 until 3) well repays a visit; it has a fine collection of porcelain made for use in the tea ceremony. The Electric Science Museum (open from 9-30 until 4, except Mondays) is of rather specialised interest. If the weather is fine, the Tsutenkaku and Osaka Towers offer extensive views over the city.

OSAKA

Location and access: Island of Honshu (west-central). Second largest city in Japan, pop. 2,800,000; capital of Osaka Prefecture, pop. 8,300,000. Distances by train (shinkansen): Tokyo 515 km (3 hr 10 min.), Nagoya 187 km (1 hr 7 min.), Kyoto 39 km (17 min.), Kobe 37 km (16 min.), Okayama 180 km (1 hr 8 min.), Hiroshima 342 km (1 hr 56 min.), Hakata 624 km (3 hrs 49 min.). Itami International Airport is 15 km from the city (20 min. by taxi). Direct flights all over the world; daily flights by ANA (All Nippon Airways) to principal cities in Japan (1 hr to Tokyo, 1 hr to Fukuoka).
Post code 530 to 536. Telephone code 06.

Travel agency: JTB, Asahi Bldg., Nakanoshima 3 chome.

Accommodation: Royal Hotel****, 2-1, Tamae cho, Kita ku, tel. 448-1121; 1,600 rooms; city centre, 3 min. from the station and 15 min. from the airport, by taxi. *Plaza Hotel*****, 2-2 Minami, Oyodo cho, Oyodo ku, tel. 453-1111; 580 rooms; central. *Osaka Grand Hotel*****, 2-22, Nakanoshima, Kita ku, tel. 202-1212; 360 rooms; central. *International Hotel****, 50, Hashizuma cho, Uchihon machi, Higashi ku, tel. 941-2661-3415; 400 rooms. *Osaka Miyako Hotel****, 110 Horikoshi cho, Tennoji ku, tel. 779-1501; 150 rooms; in the Tennoji Station building. *Tokyo Hotel****, 1-21, Toyosaki Nichi dori, Oyodo ku, tel. 372-8181; 640 rooms; close to Umeda and Shin-Osaka (shinkansen) Stations; 15 min. from the airport. *Hanshin Hotel***, 8, Umeda cho, Kita ku, tel. 344-1661; 240 rooms; opposite the station, panoramic views from the restaurant. *New Hankyu Hotel***, 38, Kobuka cho, Kita ku, tel. 372-5101; 1030 rooms; near the station and city centre. *Osaka Castle Hotel***, Temma-bashi, Higashi ku, 90 rooms; businessmen's hotel.

Restaurants: Kyomatsu, 2-2 chome, Dojima kami, Kita ku, tel. 341-0121; Japanese and Western cuisine, sukiyaki, tempura, beef specialities. *Hon Miyake,* 3-5 Dojima kami, Kita ku, tel. 341-7582; Japanese specialities. *Hon Morita,* 7, Ichihan cho, Namba Shinchi, Minami ku, tel. 211-3600; speciality: sukiyaki. *Ikutama Goten,* 39 Ikutama cho, Tennoji ku, tel. 771-1703; Japanese specialities. *Kitcho*, 3-23, Koraibashi, Higashi ku, tel. 231-1937; Japanese cuisine, luxurious. *Taikoen*, 9-10, Amijima machi, Miyakojima ku, tel. 351-8201; Western and Japanese cuisine, pleasant atmosphere. *Takoume*, 1-68 Naka, Sonezai, Kita ku, tel. 341-2861; Japanese, Chinese and Western cuisine. *Yotaro*, 32 Kasaya cho, Minami ku, tel. 211-2020; speciality: tempura. *Suehiro Asahi*, 2-40 Sonezaki Shinchi, Kita ku, tel. 341-1610; Japanese specialities: sukiyaki, shabu-shabu—which corresponds to Chinese fondue. *Senba Suehiro*, 13th story Itochu Bldg., Kitahisataro, Higashi ku, tel. 252-2140; Japanese and Western cuisines; specialities: sukiyaki and steaks.

Theatres: "Noh" theatre: Osaka No, Kaikan 12 Micimoto cho. Kabuki theatre: Shin Kabukiza, 59 Namba 5 chome. Bunraku: Asahi za 1 Higashi Yagura cho.

Festivals: 22 April, Court dances at Shintennoji; 24 and 25 July, Tenjin Matsuri, races between decorated boats, on the river; 31 August, Sumiyoshi Matsuri Shrine Festival.

OKAYAMA (end of p. 160)

cho, tel. 54-1155, 200 rooms; in the city centre, 2 min. from the station and 20 min. from the airport, by taxi. *Okayama Grand Hotel***, 2-10 Funabashi, tel. 25-1691, 30 rooms; businessmen's hotel, in city centre; 7 min. from the station and 15 min. from the airport, by taxi.
Ryokan: *Shinmatsunoe***, Ifuku cho, 39 rooms, tel. 52-5131. *Ishiyama Kadan**, Marunou chi, tel. 25-4801, 40 rooms.

Restaurants: All the main hotel restaurants serve Western food. *Koraku** (Western and Japanese cuisine), Marunou chi, tel. 22-6781.

Souvenirs: Bizen porcelain.

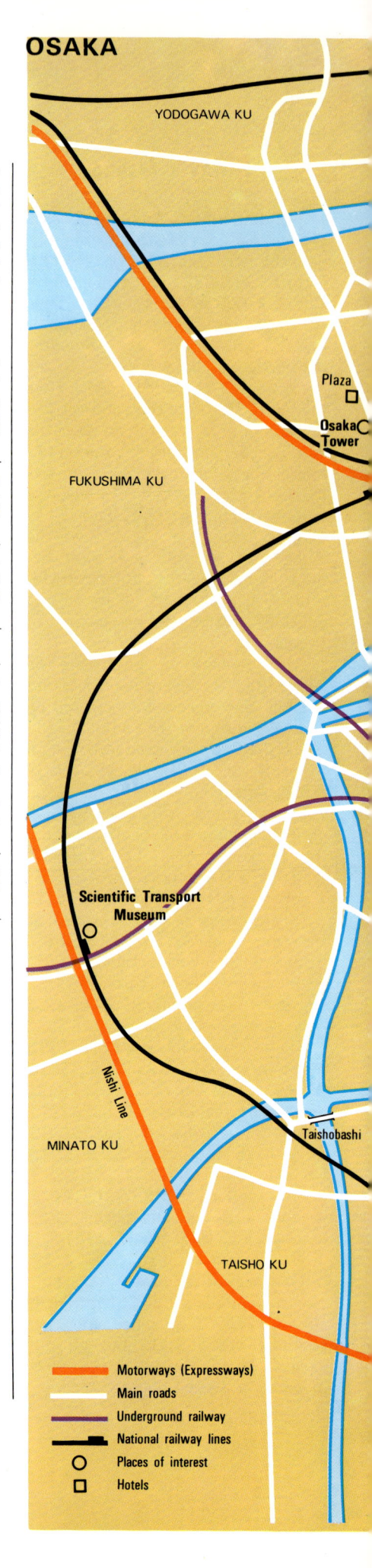

OSAKA 163

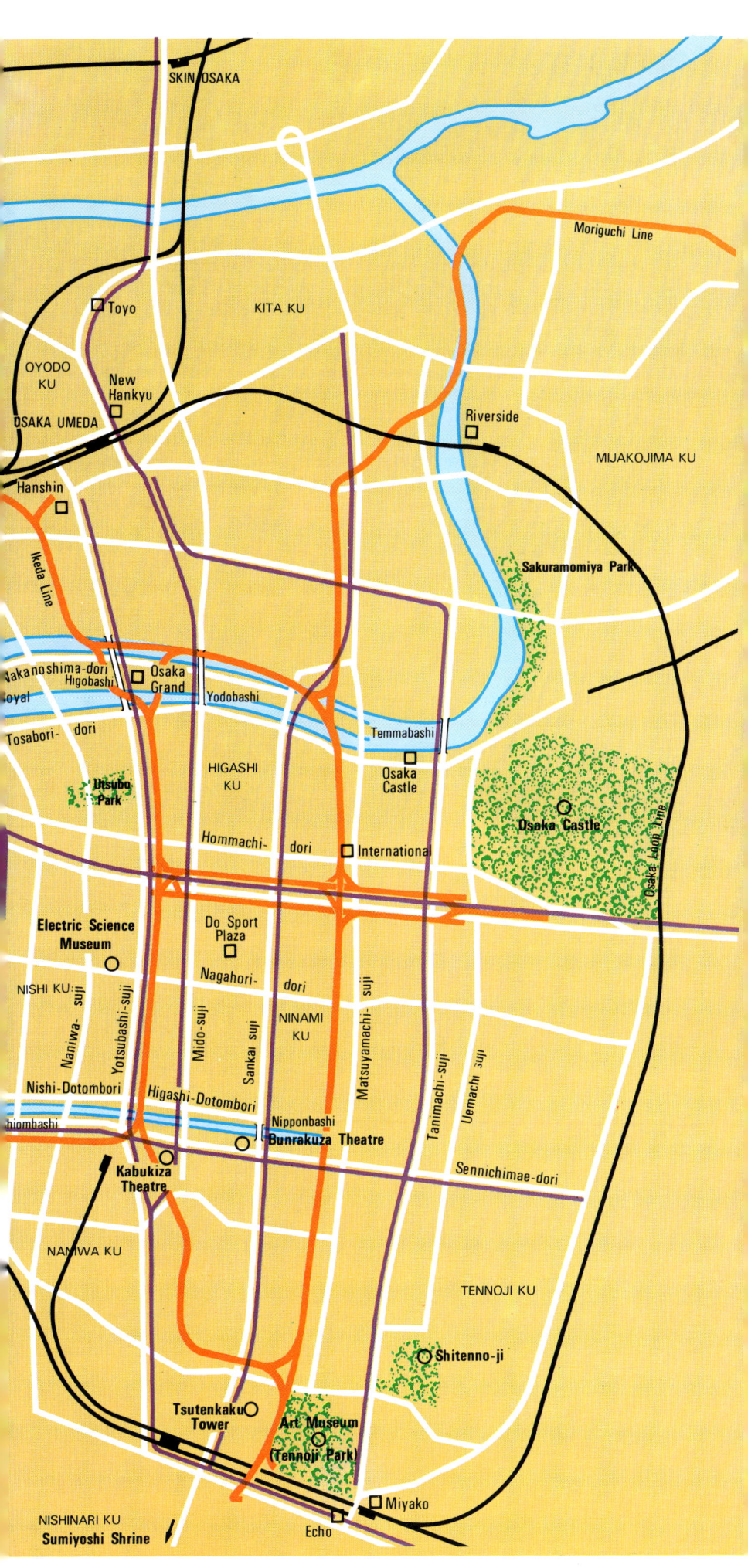

sapporo

■ If one is prepared to face the rigours of the Hokkaido winter, then the ideal time to visit Sapporo is in February, during the Snow Festival. The whole city—which is laid out like a giant chequer-board—lies covered in a thick white blanket. Children race about on sledges in the side-streets and the suburbs and the main boulevard—Odori—105 metres wide is the scene of a display of some two hundred sculptures, all made from ice and snow! Some are gigantic, some are small; they depict everything from monuments to figures to animals.

In 1975 1,750,000 visitors came to Sapporo for the Festival; in 1976 artists from eight countries took part —including some from Hong Kong and Indonesia where snow is most uncommon.

Sapporo is Hokkaido's modern administrative, economic and cultural capital. With a present day population of 1,250,000 Sato poro petsu ("great dry river") grew up around 1869 when the Hokkaido Development Commission established its headquarters here on the south-west side of the Ishikari plain. The city, planned on geometric lines with wide tree-lined streets, grew rapidly. It now has a modern underground railway. As well as the Odori Boulevard, with its lawns and fountains, dominated by the lofty television tower (fine view from the top), the liveliest parts are Nishi San Chome with its famous underground shopping gallery, and the Tanuki Koji Centre. Not far from there are the Botanical Gardens, in the middle of which is the Ainu Museum, housing a collection of Ainu craftwork as well as Hokkaido plants and animals.

Sapporo is said to be the gayest city in Japan. Winter lasts six long months so the locals do everything they can not to let it get them down! Everyone skates and skis—there are excellent facilities for winter sports, built for the Winter Olympics, on Mounts Okusa and Miyanomori. In the Susukino quarter of Sapporo there are over 3,700 bars, first class restaurants and smaller Chinese ones, not to mention quaint little cafes selling grilled fish.

There are many night clubs and pachinko parlours. Around six in the evening the streets get busy and the bars open up; in summertime you can buy ice-cream or grilled corn on the cob, or have your fortune told by a lady sitting behind a big orange lantern. At eleven peace reigns in the streets once more, the geishas have disappeared, there are just a few connoisseurs of sake and Sapporu beer to be making their way unsteadily home.

Eight kilometres from Sapporo, by tram and cable-car or (simpler) by taxi, will bring you to the top of a hill called Moiwa Yama; from here there is a good view out over the whole Sapporo plain and the nearest ski-ing station is here too. Finally, Sapporo is the starting point for excursions to the Shikotsu Toya National Park.

SAPPORO

Location and access: Island of Hokkaido (capital); 7th city of Japan, pop. 1,250,000. Airport: Chitose, with direct and frequent flights to Honshu—Tokyo, 1 hr 25 min.; Osaka, 1 hr 45 min.—as well as to the 8 other airports in Hokkaido, by TDA. Train: the journey from Tokyo takes 15 hrs 50 min. (including the ferry crossing Aomori-Hakodate). When the tunnel under the Strait of Tsugaru is completed and it is possible to do the whole journey by "shinkansen" it will take 5 hrs 50min.
Post Post Code 060. Telephone code 011.

Travel agency: JTB, c/o Nippon Seimei Bldg., 4 Kita Sanjo Nishi, Chuo ku (tel. 241-6201).

Accommodation: Sapporo Grand Hotel****, 4 Nishi, Kita-Ichijo, Chuo ku. tel. (011) 261-3311; 520 rooms; city centre; two minutes' walk from the station; 1 hr by taxi from the airport. *Sapporo Park Hotel*****, 3-11 Minami Jujo, Chuo ku, tel. (011) 511-3131; 220 rooms; the rooms at the back of the hotel have a pretty view over Nakajima Park to the Moiwa hills. *Century Royal Hotel****, 5 Nishi, Kita Gojo, Chuo ku. tel. 221-2121. 340 rooms; city centre. *Sapporo Prince Hotel****, 11 Nishi, Minami Nijo Chuo ku, tel. 231-5310; 230 rooms; city centre. *Sapporo Tokyu Hotel****, 4 Nishi, Kita Yojo, Chuo ku, tel. 231-5611; 260 rooms; city centre. *Sapporo Zennikku Hotel****, Nishi 1-chome, Kita-Sanjo Chuo ku, tel. 221-4411; 470 rooms; three minutes' walk from the station; panoramic views from 22nd floor restaurant. *Sapporo International Hotel***, 4-1 Nishi Kita Yojo, Chuo ku, tel. 261-1381; 100 rooms; opposite the station; tourist and businessmen's hotel. *Sapporo Royal Hotel***, 1 Higashi, Minami Shichijo Chuo ku, tel. 511-2121; 90 rooms; city centre.

Festival: Snow Festival on the O dori, end of January and beginning of February.

sendai

■ The largest city in northern Honshu and the most important urban centre in the Tohoku region, Sendai lies on the banks of the Hirose gawa (from which it takes its Ainu name, Sebunai, "great river") a few kilometres from the Pacific coast. A castle was built here by a local daimyo, Date Masamune, during the 17th century; his family lived there until the Meiji Restoration in 1868, but only a few ruins now remain. Fascinated by the tales of a Jesuit priest, Luis Sotelo, Masamune equipped and dispatched an expedition to Mexico, under the command of a certain Hasekura. After seven years of travelling, to Mexico, then Spain, France and Italy (where he was received by the Pope, Paul V) Hasekura returned to Sendai, a convert to Christianity and full of fascinating accounts of his adventures and of people in customs in far away lands.

Sendai is now a large industrial city but it has some fine religious buildings and lovely peaceful gardens. The Shrine of Osaki Hachiman gu, built during the 17th century, stands on the summit of a wooded hill. Beyond the two torii gateways there is an avenue of cedars set with stone lanterns. The main building, covered with black lacquer, is decorated with beautiful paintings and has an amusing carved and painted frieze of mermaids around its cornice. The little Rinno ji Temple stands hard up against the tombs of an ancient graveyard. The monastery building looks out onto an enchanting garden, laid out in the 14th century, with the traditional pools, little bridges, weeping willows and maples.

In spring the Tsutsujiga oka Park is thronged with visitors at cherry blossom time, then again a little later for the azaleas.

Sendai is a base for organised tours to Matsushima. In the mountains around the city, within a radius of up to thirty kilometres, there are many hot springs and spas, including Akiu Onsen and Sakunami Onsen.

SENDAI

Location and access: Island of Honshu (northern). Pop. 620,000; capital of Miyagi Prefecture, pop. 1,980,000. Distances by train: Tokyo 329 km (4 hr), Aomori 390 (4 ½ hr). Airport 19 km from the city, ANA flights direct to Tokyo (55 min.), Osaka (2 hr 15 min.), Sapporo (1 hr 10 min.).
Post code 980. Telephone code 0222.

Travel agency: JTB, 3-6-1 Ichiban cho, opposite the station; tel. 23-8373.

Accommodation: Sendai Hotel***, Chuo, 90 rooms (5 trad. style); near the station. *Hotel Sendai Plaza****, 2-20-1, Hon cho, tel. 62-7111, telex 852-965; 220 rooms; 5 min. from the station; 40 min. from the airport. *Grand Hotel Sendai****, 3-7-1, Ichiban cho, tel. 25-2101; 76 rooms; 3 min. from the station, 40 min. from the airport, by taxi. *Hotel Rich Sendai****, 2-2, Kobun cho, tel. 62-8811, telex 853483; 240 rooms; 5 min. from the station, 1 hr from the airport. *Hotel Koyo***, 1-7, 4-chome, Ichiban cho, tel. 62-6311; 63 rooms. *Sendai City Hotel***, 2-2-10, Chuo, tel. 23-5131; 60 rooms. *Sendai Central Hotel***, 2-1-7, Chuo, tel. 22-4161; 43 rooms. These last two hotels are close to the station and 40 min. by taxi from the airport.
Ryokan: *Miyako Hotel****, 2-9-14 Hon cho, tel. 21-3311; 37 rooms; near the station. *Akin Kokusai Hotel****, Akin Onsen, tel. 2031; 30 rooms; near the station.

Restaurant: Sendai Seiyoken* (Western cuisine), 4-175, Omachi, Shin Sendai Building, tel. 22-7834.

Youth Hostel: Sendai Akamon*, 61, Kawau chi Kawamae, tel. 21-3611; opposite the station.

Souvenirs: Wooden dolls, or "Ko keshi".

shikoku   takamatsu

■ The visitor's knowledge of the island of Shikoku is generally confined to the northern shores bordering the Inland Sea—the parts included in organised tours to Takamatsu and Ritsurin Park, Kotohira and the shrine of Kompira san.

Japanese Buddhists sometimes spend three months in Shikoku, making pilgrimages to all the island's ninety temples. Shikoku is the smallest of the larger Japanese islands (18,777 sq km); it is separated from Honshu by the Inland Sea. Pending completion of the bridge that will link it to Honshu it is easy to get to Shikoku by hovercraft—there are services from Tamano (near Okayama) to Takamatsu, and from Hiroshima to Matsuyam via Kure. There is also a boat from Matsuyama to Beppu in Kyushu, the crossing takes about six hours.

Shikoku is a mountainous wooded island and consists of two distinct regions. The northern part is attracting more and more immigrants to the large industrial cities of Matsuyama, Takamatsu and Imabari. On the broad dry plains, rice, wheat, and tobacco are grown and there are flourishing orchards. The southern part of the island is very mountainous; here, on narrow patches of flat land, rice is the traditional crop, along with early vegetables, sheltered from the typhoons by low stone walls. This whole area has a sub-tropical climate, with hot summers and wet, very mild winters.

For visits and tours, and information about accommodation in Shikoku, see the entries for Takamatsu, Koya san and Naruto as well as the chapters entitled "*Japanese Journey*".

Takamatsu (end)

Hotel Kawaoku*, Hyakken machi, tel. 21-5666, 30 rooms. There are about ten more, of all categories.

Restaurant: Yashima Sky Lounge*, (Western cuisine), Higashi machi, tel. 41-9826.

Souvenirs: The craft shop, at the entrance to the Park, has locally made goods—woven fabrics and bamboo objects.

■ Situated on the shores of the Inland Sea, Takamatsu is with Matsuyama, one of the most important cities and gateways to the island of Shikoku. Its principal attraction for the visitor is the Ritsurin koen, a large park of 78 hectares at the foot of Mount Shiun. Designed in the 17th century and formerly the property of the Mutsudaira family, Ritsurin, one of the loveliest parks in Japan, consists of two gardens, the northern and the southern. As in all classical Japanese landscape gardens there are the elements of water, islands, trees and bridges—some six pools and thirteen hills in all. Its unusual feature is its abundance of remarkable twisted pine trees, all carefully pruned to keep them small. The best time to visit is in spring, for the cherry blossom and the azaleas, and in the autumn.

About 20 kilometres east of the city (access by toll-road or funicular between Yashima Tozanguchi and Yashima Sanjo) lies the Yashima Plateau which projects into the sea and gives fine views all along the coast. This plateau became famous in Japanese history at the time of the struggle between the Heike and Genji clans (see: *Japan through the centuries* and Kyoto). In 1182, Taira Munemori, who was being pursued by Minamoto Yoshitsune, found refuge here. There is a little museum devoted to the history of the district.

TAKAMATSU

Location and access: Island of Shikoku, pop. 300,000. Distances by train; Tokyo 750 km, Matsuyama 199 km, Osaka 227 km (via Uno). Airport 5 km from city (2 hr flight to Tokyo, 40 min. to Osaka).
Post code 760. Telephone code 0878.

Accommodation: Takamatsu International *Hotel****, tel. 31-1511; 110 rooms; on the outskirts, 10min. by taxi from the station; pleasant atmosphere, quiet, good service. *Takamatsu Grand Hotel****, opposite the station, tel. 51-5757, telex 5822-557; 130 rooms; very convenient for touring; close to the harbour, to Kawaramachi Station (for Kotohira) and the stop for buses to Naruto. *Keio Plaza Hotel Takamatsu***, 5-11 Chuo cho, tel. 34-5511; 180 rooms; city centre, near business quarter, 5 min. from the station by taxi.
Ryokan: *Kiyomi Sanso****, Saiho cho, tel. 61-5580, 41 rooms. *Tadaso****, Nakashin cho, tel. 31-3176, 39 rooms. *Shinto Hotel***, Nishinchi machi, tel. 51-2559, 16 rooms.

*Luminosity and life
in big Japanese cities...*

takayama

Takayama, "a little Alpine Kyoto", is tucked away in the mountains, high up in the Miya gawa Valley. It is a most attractive small city, traversed by clear, rushing streams, with delightful streets of old-fashioned houses, the haunts of antique dealers and craftsmen of many kinds. It is rich in museums of folklore.

In winter the climate here is very severe. With its two-metre blanket of snow and bitter, crystal-clear air, Takayama then seems like an ideal setting for a film of Kawabata's famous novel "Land of Snow".

From springtime onwards, especially at week-ends, Japanese visitors flock to Takayama, attracted by its lovely setting and the contrast it affords to the heat and bustle of the big cities. Any visitor to Takayama must resign himself to being part of a crowd, but yet a happy crowd and one that will enable him to learn something about the essential nature of Japan. Like Amano Hashidate, Takayama casts a spell over those who visit it. As the little train winds its way slowly up the Miya gawa Valley the people on board gradually lose their inhibitions, picnicking happily on "sushi" (see: *Food in Japan*) and chattering away with increasing volubility and friendliness.

A full day is an absolute minimum to visit the city and two are highly advisable if its charms are to be fully explored. At the height of summer a stay of a whole week in this cool and delightful spot would be an ideal way to recuperate.

In Takayama everything is worth seeing. Rapid coach tours are available. A visitor who has little time would be well advised to take a taxi, the drivers know the standard tour very well. If you have two days to spend here, it is a good plan to devote a couple of hours to the museum in the village of Hida and then spend the rest of the time exploring the older parts of the city on foot.

The Hida minzoku mura (open from 9 until 5), to the south of the station, consists of a fine collection of regional farm buildings attractively grouped on the wooden hillside overlooking the city and around a small lake. The local inhabitants have long been expert craftsmen in wood, carpenters and joiners; their domestic architecture reflects a concern for beauty and for decoration.

Old but beautiful farmhouses

The old Arai farmhouse, a typical building of the centre of Hida province, has a wide roof covered with stone slates, supported on massive pillars, to withstand the heavy winter snowfalls. The Tanaka house, with its fine floor of rammed earth, gives a good idea of the setting of farm life five centuries ago. The Nishioka house belonged to a Buddhist priest. On the third floor there is a collection of clothes and implements used in silkworm raising. The heads of the Taguchi family were the leaders of the village over several generations. Their house is interesting for the way in which its rooms are divided, its minimal furnishing and a great central hearth and chimney. The walls of the Michine farmhouse are made of "keyaki", a rare Japanese wood and an excellent building material which the Tokugawa regime forbade the common people to use during the 17th century. The originality of the Yoshizane farmhouse lies in the fact that its two massive supporting pillars consist of two forked trees whose branches formed part of the roof. The Tomita house, an old staging post dating from the time when all goods were carried on human or horseback, contains a collection of implements used in the transport of commodities.

Close by the farm buildings there is an Arts and Crafts Centre where articles made of wood and paper, by local artists, are displayed. The Wakayama house contains the Hida Folklore Museum; this house was built in the village of Sokawa in 1751 and transported here and converted into a museum in 1959. The old furnishing of its rooms has been scrupulously preserved and they can all be visited. The Nokubi house and the Gogura barn evoke the agricultural scene during the Edo period, when the farmers' lives were a constant struggle against the elements and all the members of the community joined forces to cope with natural disasters.

After an interesting visit (though the displays are somewhat impersonal) the visitor is now free to wander through the picturesque quarters of the city. The streets named Ichino machi, Ninomachi, Sono machi and Ojin machi, give a vivid picture of "Old Japan", with their wide-roofed houses with wooden facades. Antique shops and others selling silks, kimonos, fans, umbrellas, china and lacquerwork, gleam with old gold and the muted colours of faded embroideries. Through open doorways one can catch a glimpse of a craftsman carving wood, of another painting decoration onto plates with a fine brush. The big lanterns outside the restaurants cast an inviting glow. Tempting smells filter out from shops selling sake, in front of others there are displays of decanters and glasses associated with this local speciality.

Whether it's sunny or raining the tiny narrow streets are suffused with a delightful romantic atmosphere of the Japanese past. Fine houses have been converted into museums, such as Hachiga Minzoku Bijutsu kan and Hida Minzoku Koko kan, both containing displays of craft objects; the Kyodo kan contains sculpture and traditional objects. The Kurakabe Mingei kan is a splendid old house with some rooms of grand proportions and others small and charming; it belongs to a wealthy old family and delicious tea can be taken here by visitors, in a pretty little courtyard.

At the end of the Ojin machi, in front of the Hachiman gu temple, stands a large building where the 17th and 18th-century floats used in the annual Takayama Festival are displayed. These double-storey wheeled floats, several metres high, are loaded with decoration—phoenix, birds, flowers—and lacquered and gilded; they give an impression of fantastic luxury. Yoroku Tamiguchi is said to have carved the finest, the "Kirin kai". During the festival of "Haru Matsuri", at the Hie Shrine on 14 and 15 April, these floats are a splendid feature of the procession.

The Kokubun ji Temple (near the station) is the oldest in the city. Founded in the 14th century and rebuilt during the 16th, it contains a lovely figure of Kannon, Goddess of Mercy. Beside the temple stands a three storeyed pagoda. Down by the river, 500 metres from the station, is Takayama jinya ato, a very fine building which was the residence of the Governor of the Province under the Tokugawa, during the 17th and 18th centuries. The great gateway leads into a small sober palace built of wood. Its appartments open onto a pretty garden with a traditional pool; its rice-granaries contain a collection of objects relating to the history of the city and its province.

At the end of the afternoon it is pleasant to walk up to Shiro Yama (1 km from the station), the hill where the castle once stood; it offers a good view over Takayama and the surrounding countryside. The evening can be pleasantly spent in the old quarters of the city.

TAKAYAMA

Location and access: Island of Honshu, pop. 58,000. Distances by train: Gifu 156 km, Kanazawa 180 km, Tokyo 342 km. Post Post code 506. Telephone code 0577.

Accommodation: Hida Hotel***, tel. 33-4600, 55 rooms (22 trad. style); close to the station, in the centre of the old city, mountain views. *Takayama Green Hotel****, tel. 33-5500, 100 rooms (12 trad. style); five min. from the station by car; panoramic restaurant with splendid views of the Alps. Ryokan: *Seiryu****, 6 Hachiman machi, tel. 32-0448, 22 rooms. *Takayama Kanko Hotel**, 280 Hachiman machi, tel. 32-4100, 30 rooms.

Restaurants: Susaki** (Japanese and Western cuisine) 4 Shinmei machi; tel. 32-0023. *Arisu**, (Japanese and Western cuisine) 87, Shimo-Ichino machi; tel. 32-200. *Suzume**, (Western cuisine) 24 Aioi machi; tel. 32-0300.

Festival: Takayama Festival, "Haru Matsuri", 14-15 April.

Souvenirs: sake and wooden articles.

tokyo

■ A fascinating capital where East and West, the old and the ultra-modern, exist cheek by jowl; Tokyo is a city where all races and all colours mingle in a marvellous kaleidoscope.

Many visitors think of it merely as a jumble of high-rise buildings and concrete motorways, animated by millions of busy ants—but this is a false impression, though all too easily acquired. For most visitors come to Japan on organised tours and spend only two or three days in Tokyo (often including their trips out to Nikko, Kamakura and the rest). A couple of quick city tours, often disappointing, a visit to a bar and night club—often chosen specially for "tourist appeal", usually leave the impression that Tokyo is an entirely modern city that keeps up a few "traditional spots" just to amuse the fleeting visitor.

Tokyo is a vast and complex place that simply cannot be "done" in a couple of days; it needs to be seen and assimilated gradually. To know it you have to explore it on foot, get off the modern thoroughfares and roam through working class districts such as Asakusa, along streets lined with smart shops and restaurants, through the many bazaars, the quiet modern residential areas and those parts where, scarcely untouched by the 20th century, little wooden houses still stand amid their tiny immaculate gardens.

You need to be exhausted by Tokyo to start with, only then will you begin to feel its spell; saturate yourself in its diversity! Tokyo is a vast conglomerate of varied buildings, peoples and civilisations.

See Tokyo on a Sunday if you can. Then the Tokyoites shed for a while most of their Western hustle and live, for a brief day, at a more traditional pace. The twelve-storey buildings don't seem to matter any more. The crowds in the streets are relaxed; they stroll along, young and old, wearing kimonos or Western dress, stopping for a drink or a meal, to have their fortunes told or perhaps to buy crickets in a cage. Watch them as they wander round the many beautiful gardens of the city—the fascination with which they gaze at each little twisted tree or lovely view! Sunday is a great day for visits to temples and shrines, the Japanese flock to them in their millions—often wearing traditional dress—to be married, to present their new-born babies, or just to stroll about and show a sign of life and interest to their gods.

Tokyo is the ideal city for the attentive visitor to get a true idea of the essence of Japanese life. It is here that he can pick up the essential clues that will show him how the Japanese achieve their extraordinary flexibility of attitudes, how they are able to pass from one extreme to another and take the best from a whole variety of situations.

Tokyo in four days

The following are a few suggestions for a comprehensive four days' sightseeing in Tokyo:

First day. Morning: the Imperial Palace.
Afternoon and early evening: Asakusa (the Temple and the district).
Second day. Morning: the Meiji Shrine.
Afternoon: the Ginza district and one of the big stores.
Third day. Morning: the National Museum (have lunch there).
Afternoon: Koraku-en Gardens, then, around 6 o'clock, the Kanda "Latin Quarter".
Fourth day. Morning: Tokyo Tower and Sengaku ji Temple, the tombs of the 47 Ronin.
Afternoon: a visit to a Kabuki Theatre (book seats in advance through your hotel).

An overall view of Tokyo from the Tower

The Tokyo Tower, a graceless metal construction 332 metres high, was built as a television tower in Shiba Park in 1958. It is the tallest

*Tokyo is a city of many faces,
where East and West live cheek by jowl
and the ultra-modern jostles the traditional
—in both buildings and life-styles.*

metal tower in the world. At its base stands the ultramodern Prince Hotel and the Zojo ji, temple of the Jodo Buddhists. The temple has a fine two-storeyed gateway, dating from 1605. On the frist platform of the Tower there is a Science Museum as well as shops and restaurants. There is another viewing platform above it, at 232 metres. When the weather is fine it affords an extensive view of Tokyo and its suburbs.

The city centre

Over the centuries Tokyo has expanded outwards, in concentric circles, from the Imperial Palace. The latter remains the city's nucleus but its urban heart is now slightly to the east, in the Chiyoda ku and the Chuo ku (ku = district).

The Chiyoda ku comprises, as well as the Imperial Palace, the business and commercial district known as Marunouchi; here there are banks, the Kokusai Building which houses the Imperial Theatre and the Idemitsu Art Gallery (open from 10 until 5). This gallery has displays of fine Japanese ceramics and prints by Hokusai, Utamaro, Hiroshige and Moronobu. Following the palace moat to the south you come to Hibaya Park; this is a pretty Western-style public garden, noted for its displays of azealeas and chrysanthemums in May and October. Not far from here, on a little hill, stands the Hie Shrine, a popular place during the Edo period; the buildings were rebuilt after the war. A great festival, with a procession with palanquins, takes place at this shrine on 15 June. City guides never fail to point out the Diet Building, a concrete structure faced with Okinawa granite, built between 1918 and 1930.

North east of the Imperial Palace lies Kanda, a sort of "Latin Quarter", humming with life. Here are many schools and colleges, universities and publishing houses. It's pleasant to stroll along the "Street of Booksellers" (Jimbo cho) and in the little lanes nearby where there are some three hundred bookstalls. It is picturesque by lantern light in the evening, with crowds of all ages jostling round the stalls that overflow onto the pavements and even the roadways. In the northern part of Kanda is the Transportation Museum (see paragraph: "Tokyo's Museums").

The Chuo ku district lies to the east of Chiyoda ku, on each side of the famous Nihombashi Bridge; from this bridge, rebuilt in stone in 1911 to replace an earlier wooden structure, all distances from Tokyo are measured. To the south lies the wellknown Ginza district (see paragraph: "Ginza"), the Kabuki Theatre (see "The Kabuki Theatre") and the Bridgestone Gallery (see "Tokyo's Museums").

Tokyo's residential areas

On each side of the city centre lies a very different residential quarter.

Shitamachi, to the east, on the banks of the Sumida river, is a working class area with a population of a million; it preserves something of the character of "old" Edo. Its inhabitants work hard and play hard —they drink plenty of sake and are enthusiastic supporters of "sumo" wrestling (see: Sport) and the Kabuki Theatre. The northern part of Shitomachi includes the extensive Lake Shinobazu (2 km in circumference), famous for its marine animals. Tokyo Bay used to extend inland as far as this lake. In the middle there is a little island on which stands a tiny temple to Benten, Goddess of Fortune. Ueno Park overlooks the lake. Here amongst the trees is the Tokyo Metropolitan Art Museum, the National Science Museum and the National Museum of Western Art (see: Tokyo's Museums). The nearby Toshogu Shrine, dedicated to Tokugawa Iyeyasu, dates from the 17th century; it is surrounded by fifty bronze lanterns.

East of Ueno Park is the picturesque Asakusa district (see paragraph: Asakusa), the Sumida area and the Kiyosumi Gardens (open from 9 until 5-30). The latter is a classic Japanese landscape garden, with little pools filled with huge carp, arched bridges, lanterns, dwarf pines and fine plantings of rhododendrons.

Yamanote, the second residential area, lies west of the Shimbashi district and includes Akasaka, Roppongi, Shibuya and Shinjuku. Yamanote is full of gardens, hills, valleys and steps and used to be the military stronghold. Today the area is essentially middle class, a place where civil servants, teachers and doctors live, either in high-rise flats or in little wooden houses set in tiny gardens, lost amid a network of narrow lanes. Such peaceful residential areas often lie only a few metres from busy commercial arteries.

Yamanote includes all the Shiba Park district (Tokyo Tower) and extends south to the Sengaku ji Temple, founded in 1610. To the left of the entrance gate are the tombs of the 47 Ronin whose dramatic story inspired a successful Kabuki play (see special entry). The fountain under which Kira's head was washed still stands in the temple courtyard.

The Roppongi and Akasaka districts form Tokyo's diplomatic quarter. The Togu Palace, residence of the Crown Prince, is here too, surrounded by a huge park, and the Akasaka Detached Palace as well—a late nineteenth-century building this, built of different coloured granite and marble specially imported from Europe. This latter palace is used by distinguished foreign visitors. The park is the setting for an annual "Chrysanthemum Viewing Ceremony", in the autumn, when the Emperor and his family come here to admire the blooms; many Tokyoites follow their example.

To the east of the gardens lie the Shibuya and Shinjuku districts, particularly lively in the evening with their restaurants, theatres, patchinko parlours, bars and night clubs. On the square in front of the Shibuya underground station stands a statue

"THE HOUR OF THE MONKEY"

At the death of the Buddha, the Paranirvana, twelve animals came running: the rat, the bull, the tiger, the rabbit, the dragon, the snake, the horse, the sheep, the monkey, the cock, the dog and the boar. These animals form the Sino-Japanese calendar, which works on a twelve-year cycle. This calendar, together with another astrological calendar in which five "elements" are represented by planets: Mars for fire, Saturn for earth, Venus for metal, Mercury for water and Jupiter for wood, gives rise to a complex art of divining the character of an unborn child, the choice of a spouse or the date of a ceremony. Animals are also used to designate the cardinal points: the rat for the north, the cock for the west, the horse for the south and the rabbit for the east. In the countryside, to this day, the hours are similarly named; thus one may have an appointment at the "hour of the monkey" (4 pm), or go to bed at the "hour of the dog" (10 pm).

of a dog called Chuken hachiko; this faithful animal waited here for years (he was fed by kindly passers by), in vain, for his master to return from the war.

The Outer Garden of the Meiji Shrine contains the Olympic Park, the Olympic Stadium (seating 80,000) and the Olympic Pool—all built for the Games in 1964. The Meiji Shrine itself is a bit further west, in the Inner Shrine Gardens. The Shinjuku gyoen Gardens lie slightly north east of the Shrine (open from 9 until 4). These lovely gardens contain a collection of rare varieties of cherry tree which are a great attraction when they are in flower. In autumn an important chrysanthemum exhibition is held here.

The most beautiful of all Tokyo's gardens are those of Koraku en. They are to the north of the city, in the Bunkyo ku district, which also contains the Todai or National University; many students live in this neighbourhood. Koraku en (open from 9 until 4-30) was laid out during the 17th century, for a member of the Tokugawa family, by the Chinese scholar Shu San Sui. A fine Chinese gateway, the Kara mon, marks the entrance. The so-called "Bridge of the Full Moon" leads across to a little island in the lake, where stands another temple dedicated to the Goddess Benten.

The Imperial Palace

Twice a year, on 2 January and 19 April, the Emperor's birthdays, the gates of the Imperial Palace Gardens are opened and the crowds come in to applaud the Emperor.

The Imperial Palace lies at the very heart of the city of Tokyo. Its buildings are enclosed within a vast park of fields, woods, pools and hills; they are almost invisible from outside. The whole Imperial domain is enclosed by fifteen kilometres of moats.

In 1457, a lord named Ota Dokan built a castle here. The place was known as Edo then, from its position on the river Sumida (Edo = "gate to the estuary"); its only inhabitants were a few fishermen. Ota's castle was surrounded by high walls with gates cut into them; a great avenue of pines led down to the sea.

When, during the 17th century, the castle became the property of Shogun Tokugawa Iyeyasu he had the whole thing pulled down and started to build a huge fortress surrounded by ditches reinforced with great granite walls. As there was little stone available in Edo, Iyeyasu asked his daimyo (vassals) to contribute to the building of his castle by sending blocks of stone quarried in their own provinces, especially in Izu. Three thousand ships were fitted out to transport the stone; the job took ten years. At their own request certain workers were buried alive in the foundations as "human pillars"; they considered this sacrifice as an honour pleasing to the gods. Following the earthquake of 1923 a number of skeletons were discovered.

Edo became the Shoguns' capital. At the Meiji Restoration the Castle passed to the Emperor who set up his Court there. In 1875 the buildings were severely damaged by fire, but they were scrupulously restored. They suffered again, from bombing, during the Second World War, but have now been restored once more. The public is not normally admitted to the Imperial Palace. A great esplanade, separated from the Hibiya Park and the Marunouchi district by the Hibiyabori and Babasakibori Moats, is a favourite promenade for Tokyoites. It is crossed by the Uchibori dori Avenue, from north to south. The Nijubashi Gateway and the Ishibashi Bridge, used by distinguished visitors, separate the esplanade from the Imperial Palace grounds. The gateway is opened to the public twice a year. This spot, overlooked by the graceful Fushimi tower, is a favourite place for provincial visitors to Tokyo to have themselves photographed. The whole esplanade, shaded by willows, following the moats that charmingly reflect the gateways and pavilions of the palace, is a wonderfully serene and peaceful oasis, only a few hundred

*Amaterasu, the Shinto Sun Goddess,
hid herself in a cave,
plunging Japan into darkness.
Her brilliant re-emergence
was the origin of the "matsuri" festivals.*

metres from the throbbing Marunouchi business quarter.

North-east of the Imperial Palace lie the Higashi gyoen Gardens. They were laid out, on various levels, in 1630, by the celebrated landscape architect Kobori Enshu; they were reconstructed after the Second World War.

The Ginza district, a shopper's paradise

The Ginza is losing its charm, the "fubutsushi" is fast fading, say the Tokyoites. The noisy tramcars have disappeared, the weeping willows have been dug up (they were its emblem), even the "Nichigoki"—the famous dance theatre that produced many stars—has closed down! But the Chuo dori Boulevard remains *the* shopping street in all Tokyo and on Sundays it becomes a pedestrian paradise as cars are banned there then.

Weekdays may be a bit grim, with vast hurrying crowds, but Sunday is the time when people go to the Ginza just for the pleasure of "gin-bura" i.e. strolling idly along. There they are, provincials, peasants, women in kimonos or in Western dress, young people window-shopping others queueing at open air stalls where biscuits and other nice things are cooked while you wait, then crowding into the big stores or little Chinese and Japanese restaurants. Ginza is derived from "gin" (money) and "za" (seat). It was here that Shogun Tokugawa Iyeyasu is said to have established the Mint, in 1612. Ginza remains a district of jewellers —their shop windows filled with pearls—of photographic dealers, as well as big stores such as Daimaru, Matsuya, Matsuzakaya and Takashimaya.

A big store

The huge facade of the big store rises from the Chuo dori, a carbon copy of its neighbours. The crowds pour in through the doorways, between a double row of uniformed (and gloved) hostesses who bow and smile in welcome. On every floor there are further relays of hostesses to welcome and assist the customers.

Half an hour before opening time the departmental heads will have had a meeting with their hostesses and saleswomen to tell them about new items and goods to be specially featured today. They will have told them yet again that their object is not to force people to buy, but to attract them into the store and retain their interest by telling them what goods are on offer in Japan and elsewhere, and to give them a few new ideas.

Their aim is to encourage the crowds that stroll through the Ginza district on Sunday to approach the big stores as they would a cinema or a theatre, though here they see not a show but a kaleidoscope of goods of all kinds.

The windows would be the despair of a Western window-dresser—goods are shown without any concern for artistic presentation; the windows seem to exist merely to give the passer by a tantalising glimpse of the inside of the store (they are not screened off behind) so that he will be tempted to go inside.

The big stores in the Ginza sell everything, from buttons to Old Master paintings, from food to kimonos. Goods are displayed without the slightest apparent concern for stylish display. All prices are clearly marked so the customer has no need to ask what things cost. The clothing floor is divided into different areas; the cheaper clothes are set out in the middle, with little "boutiques" selling well-known European brands of clothes and jewellery all around. Every big store prides itself on having an elegant "Courrèges", "Dior" and "Gucci" boutique.

The kimono department retains a traditonal atmosphere. There are showcases with splendid white or jade-green kimonos, plain or elaborately patterned. There are tables piled high with bolts of cloth for kimonos or obis (the wide belts worn with the kimono). Customers stand on "tatami" (mats) to have their measurements taken or to try garments on. Only 18% of Japanese families buy a kimono every year, for even a simple one costs 50,000 yen; the most elaborate can run into millions. Visitors from abroad will be surprised to see such a large department specialising in golfing accessories —golf is a fairly recent craze among wealthy Japanese. The variety of tableware is astonishing too. Tables groan under piles of china in blues and greens and browns. Rarer pieces—sake cups and tea bowls—are carefully arranged on special stands.

A whole floor is devoted to paintings including Old Masters, to rare plants (certain bonsai cost as much as 200,000 yen), and to exotic fish (a venerable carp may be priced at a million yen). Every big store has its marriage department too.

The aim of the big store is to please its customers. They can spend hours wandering round the various luxurious departments, then watch a kabuki play, perhaps, or some bunraku marionnettes, or perhaps see a film. The theatre and the cinema are usually on the top floor. Stores organise all sorts of exhibitions, of swords for example, or old prints. Thus a family can easily spend an entire day in the store. Traditional and Chinese restaurants offer appetising meals. On the ground floor there is often a so-called "French Bar" where Italian coffee and pastries are served, on a pavement terrasse as well, if the weather is good enough.

The basement food department is particularly fascinating for a foreign visitor. You feel as if you are in a market; meat counters are covered with meat of various qualities, finely sliced for "sukiyaki" (see special entry). Meat tends to be very expensive in Japan: fillet of beef costs around 20,000 yen a kilo, mince about 5,000; pork is a little cheaper, about 3,000 yen a kilo.

At the fish counters you can learn how the Japanese prepare fish for

THOSE TORTURED DWARF TREES: "BONSAI"

■ You see "bonsai"—dwarf trees, anything from ten to a hundred years old—everywhere in Japan; in private houses, in exhibitions in big stores or even in railway stations; the Japanese are devoted to them. This is how they are grown.

The seeds are planted in tiny pots. In due course they germinate and grow. They are left in their little pot until their roots fill it completely. Then they are transplanted into a pot just marginally larger than the first—and so on throughout their lives. The trees are literally starved and their root development is restricted, so they never grow to anything like their normal size. The branches are tied together an and to the trunk, with bamboo thongs; thus their growth is constricted and slowed down. In this way twisted "ancient" forms are produced, in trees no more than 50 centimetres tall. In public parks and gardens all new shoots on pine trees are frequently removed by hand, in order to prevent them from growing. Conifers are the species that lend themselves most readily to this form of vegetable martyrdom, but oaks and beeches are often treated as well.

TOKYO: MAIN ACCESS ROADS AND ADMINISTRATIVE BOUNDARIES

178 TOKYO

Tokyo

Hotels

By category and in alphabetical order

★★★★★
1. Imperial Hotel
2. The New Otani

★★★★
3. Akasaka Tokyu Hotel
4. Ginza Tokyu Hotel
5. Keio Plaza Intercontinental Hotel
6. Hotel Okura
7. Hotel Pacific
8. Takanawa Prince Hotel
9. Tokyo Hilton Hotel
10. Tokyo Prince Hotel

★★★
11. Akasaka Prince Hotel
12. Azabu Prince Hotel
13. Ginza Dai-Ichi Hotel
14. Hotel Grand Palace
15. Haneda Tokyo Hotel Air Terminal
16. Marunouchi Hotel
17. Hotel New Japan
18. Hotel Palace
19. Shiba Park Hotel
20. Shimbashi Dai-Ichi Hotel
21. Fairmont Hotel Tokyo

★★
22. Diamond Hotel
23. Gajoen Kanko Hotel
24. Ginza Nikko Hotel
25. Hill Top Yamanoue Hotel
26. Hotel Kokussai Kanko
27. Hotel Takanawa
28. Takanawa Tobu Hotel
29. Takara Hotel
30. Tokyo Air Terminal Hotel
31. Tokyo Hotel Urashima
32. Tokyo Kanko Hotel

★
33. Hotel Daiei
34. Hotel New Meguro
35. San Bancho Hotel
36. Hotel Tokyo
37. Hotel Toshi Center

A tour of the city and its monuments

1. Imperial Palace
2. Marunouchi: the business area
3. Yurako-cho: cosmopolitan area (JNR station)
4. Asahi Shimbun (newspaper) building
5. Imperial Hotel
6. Tokyo Station
7. Kanda: Tokyo's 'Latin Quarter
8. JTB Transport Museum
9. Science Museum
10. National Museum of Modern Art
11. Akasaka: busiest shopping and entertainment area
12. The Diet Building
13. Kasumgaseki: administrative area
14. Hibiya Koen, park and municipal library
15. Ginza: one of the most colourful areas
16. Kabuki-za Theatre
17. Chuo Dori: street of famous big stores
    a) Matsuzakaya,
    b) Mitsukoshi,
    c) Matsuya,
    d) Takashimaya,
    e) Daimaru,
    f) Sogo,
    g) Komatsu
18. Bridgestone Gallery
19. Nihon Bashi: banking area
    (Bank of Japan and Stock Exchange)
20. Central Market
21. Hama Rikyu Park
22. Shimbashi: busy colourful area
    (JNR station and underground)
23. Tokyo Tower
24. Sengaku ji, Buddhist temple
25. Okura Museum
26. Roppongi: residential and night life area
27. Nezu Art Museum
28. Aoyama Palace
29. Akasaka Detached Palace
30. Shinjuku: busy colourful area
    (JNR station and underground)
31. Shibuya: busy colourful area
32. Koraku en, one of the loveliest gardens
33. University
34. Lake Shinobazu
35. Kiyomizu do Temple
36. Tosho gu Shrine
37. Tokyo National Museum
38. National Science Museum
39. National Museum of Western Art
40. Asakusa Kannon Temple

"sashimi" (see: Japanese cooking), according to their quality. There are baskets heaped high with seaweeds, green, yellow, brown and black, wide and narrow, thin and thick. At the tea counter, customers are busy tasting many different kinds. Fruit, all carefully wrapped, is dear—an apple costs 500 yen. Little white banners flutter over everything, markers indicating the prices.

In Tokyo, as in all the provincial cities, it is the big stores that really set the fashions in most things. Statistics reveal that their customers spend a minimum of 3,000 yen each, per visit.

The Goddess Kannon watches over Asakusa

After passing under an huge gateway, a long avenue lined with stalls selling souvenirs, kimonos, and things to eat leads the visitor to a temple whose origins go back to the 7th century. Three fishermen are said to have founded the Asakusa Temple to shelter a little gold statue of Kannon, Goddess of Mercy, that they found one morning in their net. In those days Asakusa was a little island, famous for the delicious seaweeds harvested in the neighbouring marshes. The main temple building, the Kannon do, the gateway and the five-storeyed pagoda dating from 1651, were all burned down during the war and have been rebuilt. The ceilings of the great hall are decorated with paintings—"Dragons" and "A Rain of Flowers from the Gods"—the work of contemporary artists.

The temple is crowded with visitors on Sundays. Young and old, they come for purification amid the smoke of hundreds of incense sticks rising slowly from a gigantic cauldron.

Then the pilgrims climb the steps to the temple itself, clapping their

"TEIKA", A NOH PLAY

■ *During the 12th century a certain Princess Shikishi and a Fujiwara Minister, an amateur poet who wrote under the name of Teika, are said to have fallen in love...*
A monk, arriving in Kyoto and caught in a sudden shower of rain, seeks shelter in a deserted house. A woman takes him to see the grave of Princess Shikishi which happens to be close by. A vine climbing over the tombstone is a symbol of Teika's undying love. This continuing passion is tormenting the Princess, lying in her tomb, and preventing her from attaining the state of "deliverance" or Buddha-hood. The woman confesses to the monk that she herself is the spirit of the Princess; she asks for the monk's help and the support of his prayers. Whereupon she disappears. The monk, filled with compassion, reads some passages from the Sutras. The tomb opens, revealing the Princess, sumptuously dressed. She is filled with gratitude to the monk, for hope has now been restored to her...

hands to summon the gods, then they make their offerings, spend a moment or two in meditation and then go off to enjoy themselves elsewhere in Asakusa.

A foreign visitor can easily spend hours in Asakusa, strolling along the narrow streets, under the covered arcades with their picturesque shops selling china, food, kimonos, combs and wigs, not to mention Chinese remedies such as snakeskins, powders and roots. In a corner nearby there is a little Shinto or Buddhist altar with incense rising lazily before it... A showcase with good luck charms—a "Daruma" and a "Maneki neko"—attracts prospective customers.

In the evening it's the old Edo that comes to life here in Asakusa. From time immemorial it has been a pleasure quarter. In feudal times people flocked here from a long way away to pray before the statuette of the Goddess Kannon, said to possess miraculous powers. After their devotions the pilgrims usually wanted a little amusement before returning home, so many little shops grew up here, some selling prints, others housing a total of two thousand courtesans of high and low degree! This licentious quarter was often destroyed by earthquake or by fire but it was always rebuilt. During the 18th century it became a favourite subject for the painters of the "Ukiyo e" school (see: *Arts*). After the war the notorious houses were closed down. Asakusa remains a popular pleasure quarter today.

People come here after work to do their shopping in this vast bazaar. From 5 o'clock on you can see the regular customers pushing back the coloured curtains that hang in front of the restaurant doors and sitting themselves down in front of bowls of "soba" or plates of "sushi" (see: *Japanese Food*). Often, dressed in their yukata (cotton kimono), they will make their way to the gaming houses; the streets are alive with the

MOUNT FUJI

Ever since earth and heaven
Were divided
In Suruga
The great peak of Fuji
Stands sublime,
Like a god.
When I see it from afar
In the plain that is the sky
It hides the rays of the sun
Beaming through the heavens,
The shining moon
Is hidden by it
The white clouds
Hesitate to pass.
The snow falls there
Without concern for season.
It will be always on men's lips
This great peak of Fuji.

YAMABE NO AKAHITO
(first half of the 8th century)

Mount Fuji epitomises youth, its outlines suggest movement, elan; it is supple, proud as a sword, an invitation to deeds of courage; it connotes chastity, virginity and youth—it's no accident that it is held to be the dwelling place of the "Princess who makes the trees blossom". Mount Fuji is a constant presence, in love and death and in all the great dramas of life. Fuji is the "Sea of Trees", the "Ju-kai", the wild pines that cover its northern slopes, Fuji is ashes, Fuji is snow. Fuji, like poetry, aspires to the heavens, and one is never quite sure whether it really belongs to this world at all; perhaps it is a "mysterious kami".

FOSCO MARAINI

TOKYO — Metropolitan area — hotels and points of interest

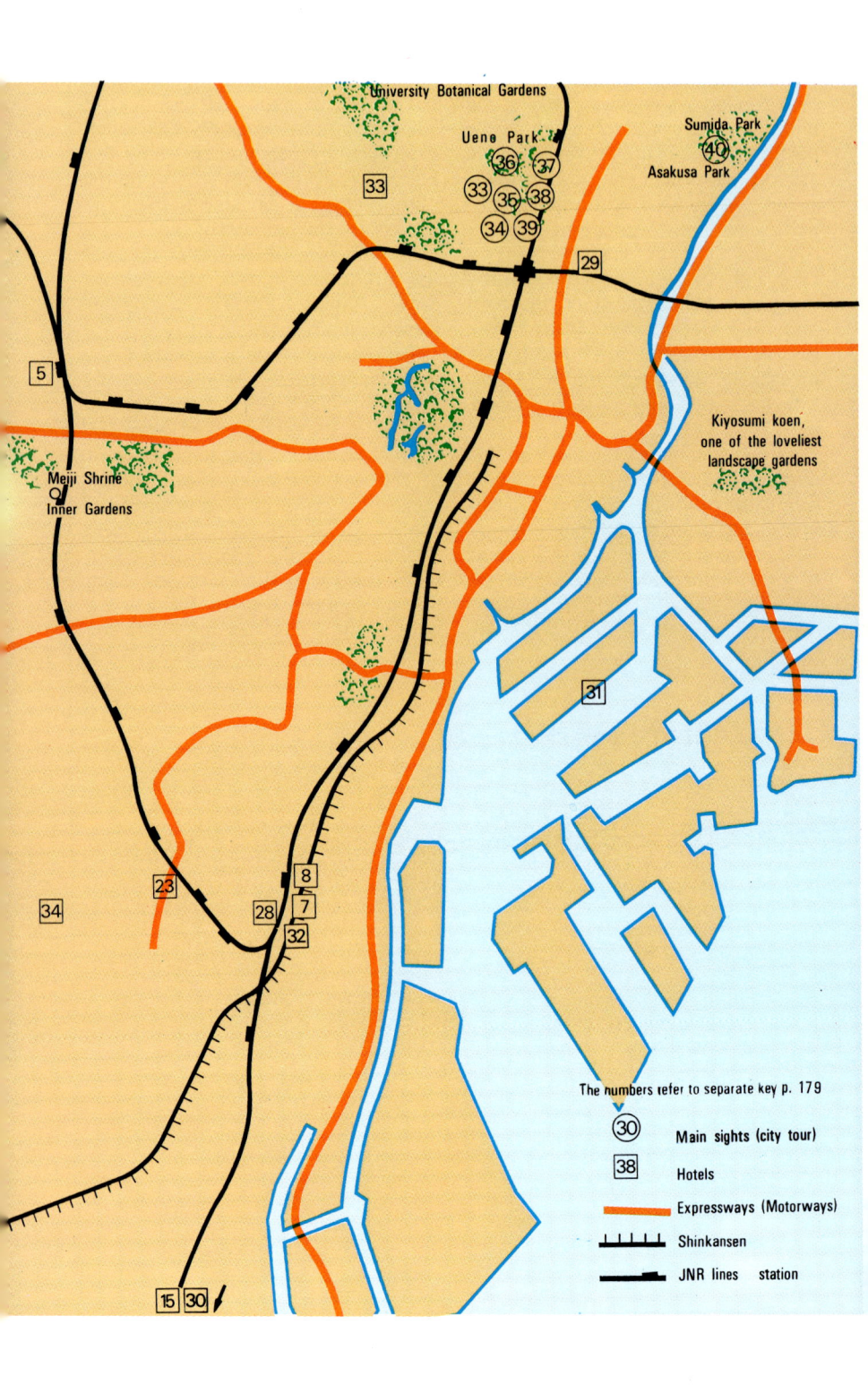

182 TOKYO

TOKYO — City centre — hotels and points of interest

click of dominoes and Mah-jong pieces. Patchinko parlours too, attract an enormous number. Snatches of music waft out from the dance halls. Geishas trot along the streets, intent on keeping their next appointment. Life goes on at a hectic pace here until late at night. When the shutters are finally put down, on the almost deserted streets, there will usually be a few figures staggering along, slightly the worse for sake, on their way home at last.

The Meiji Shrine and its Inner Garden

In the centre of Tokyo lies a huge park in which stands the Meiji Shrine. The so-called Inner Garden, planted with some 130,000 great trees—offerings from the various provinces of Japan, is an ideal place in which to walk and take the air. The Shrine itself is dedicated to the Emperor Meiji and his wife. It is reached down wide avenues set with great torii (gateways), made of hinoki wood—from ancient cypress trees, imported from Formosa. The Emperor Meiji was much esteemed as an enlightened ruler and on his death, in 1912, the Japanese people decided to erect a great monument in his honour. The Shrine, completed in 1920, was badly damaged during the war but has been well restored.

A Sunday visit is much to be recommended. The gardens and the shrine are thronged with colourful crowds, among them many young couples who are coming to present their new-born babies at the shrine, in a ceremony known as "Miya mairi". The babies are all swathed in richly embroidered robes and their mothers wear splendid kimonos.

In the northern part of the park stands the Homotsu den Museum (open from 9 until 4-30). It contains objects belonging to Emperor Meiji,

HOW TO COMMIT "SEPPUKU", OR "HARAKIRI"

■ *This used to be the honourable way of committing suicide in Japan. A man who had to, or wanted to, commit suicide would nominate an assistant, a friend or relation, who would consider it a great honour to be asked to perform such an important last service. He would then put on the appropriate ceremonial kimono, go to a specially designated pavilion, greet any witnesses, and sit down in front of a tray bearing a nine-inch dagger, wrapped in white paper so that only the point of it showed. The suicide would then grasp the dagger by the middle of the blade and plunge it into his stomach, cutting across horizontally. Then his friend would immediately decapitate him, with one blow of his own sword. Usually the "second" had asked permission to follow his friend into death and would then commit "seppuku" in his turn, helped by another, who, in due course... the chain of suicides would finally be brought to an end by the last unfortunate committing "seppuku" unaided.*

including a carriage used at the proclamation of the Constitution, in 1889.

South of the shrine there is the most beautiful iris garden in all Japan, planted with over a hundred and fifty different varieties of the flower. The ideal time to see it is at the end of June and beginning of July, when there are many waterlilies in flower there as well.

Tokyo's Museums

It is simply not possible for a visitor with only a few days at his disposal to see anything like all of the artistic treasures on display in the museums of Tokyo. He has to limit his choices; what follows is a possible selection.

The National Museum (open from 9 until 4-30, except Mondays) covers the whole history of Japan and Japanese Art, from the earliest periods. Its collections of archaeology, sculpture, painting, lacquerwork, ceramics and armour, make it the most complete and most interesting museum in the whole country.

Built before the Second World War, it is housed in a number of buildings. Some 3,000 of its collection of 40,000 items are displayed in rotation.

The ground floor of the main building, "Hon kan", (facing you as you enter), contains sculpture, swords, armour, porcelain, fabrics and kimonos. After the sculpture of the Nara and Heian periods there is a fine Kamakura period (14th century) figure of Minamoto Yoritomo, in polychrome wood with eyes of rock crystal.

Another room contains Noh and Kyogen masks (see: *Theatre*); elsewhere there are collections of metalwork—daggers, swords and swordguards, and armour—and porcelain, including many pieces made for use in the tea ceremony.

THE SOROBAN OR JAPANESE ABACUS

■ *A visitor to Japan, buying pocket calculating machines in a specialist shop, may well be astonished to see the assistant quickly adding up his bill on an abacus. Japan, known the world over for her gadgets and computers, remains faithful to the abacus, a Mesopotamian invention introduced from China in the 16th century. The "soroban" consists of a rectangular wooden frame, divided into two unequal parts by a transverse bar, with rows of metal rods with balls threaded on them.*
On one side of the bar each rod has five balls, each counting as one unit; on the other side of the bar each ball counts as five units. This simple calculating machine is still extremely popular in Japan. With it is possible to do addition and subtraction of even large numbers more quickly than with a complicated computer. The teaching of the use of the abacus is compulsory in primary schools and there are some 1,200 private abacus schools in Tokyo and the same number in Osaka. Candidates for examinations set by the Japanese Chamber of Commerce and Industry have to take a certificate in proficiency with the abacus. They are extensively used in banks; they are cheap (about 1,000 yen each), convenient, and easy to use. Office workers frequently share the use of an electronic calculator but have an abacus of their own for simpler calculations. A few years ago an abacus champion took on a computer in a series of simple and complex calculations—the man with the abacus produced his answers first!

The First Floor is devoted to painting, calligraphy, lacquerwork and woodwork. All schools of Japanese painting prior to the twentieth century are represented.

From the Heian period (794-1185) note the 12th-century "Fugen Bosatsu", a painted scroll depicting the Buddha seated on a white elephant. From the Kamakura period (1192-1333) there is a display of "emakimono" (see: *Art*): the Jigoku Zoshi' (hell scroll), the "Gaki Zoshi" (scroll of starving spectres), whilst the "Ippen Shonin Eden" depicts the pilgrimages of the monk Ippen. The finest work of the Muromachi period (1337-1573) on display is by Sesschu: "The Three Landscapes". In the Momoyama period (1586-1615) the Tosa school is represented by Tosa Mitsuyoshi's "Landscape by Moonlight", and the Kano school by Kano Eikoku's "Pinewood" and "Trees and Clouds and Mountains" and "Tchao fu and the Ox". In the Edo period (1615-1867) there are paintings of the Ukiyo e school, as well as many prints by Hokusai, Utamara and Hiroshige (famous Kabuki actors and courtesans). In one room there is a display of lacquerwork. The history of calligraphy is traced, in another, from its Chinese origins through the Heian and Kamakura periods.

To the right of the Hon kan stands the Toyo kan, a modern building containing Eastern works of art that have influenced the arts in Japan itself. To the left of the Hon kan is the Hyokei kan, which houses Japanese archaeological collections, notably Haniwa objects (see: Art and the entry for Miyazaki).

So many museums

Also in Ueno Park, not far from the National Museum, is the National Science Museum (open from 9 until 4-30, except Mondays). Its collections cover zoology, botany, meteorology, geography physics and chemistry.

The National Museum of Western Art (open from 9-30 until 4-30, except Mondays) displays fine Western paintings and sculptures from the collection of Matsukata Kojiro, including works by Rodin, Bourdelle, Cézanne, Courbet, Gaugin, Monet and Renoir.

The Okura Museum, 3 Aio cho, Akasaka Minato ku, (open from 10 until 4, except Mondays) has a collection of Japanese, Chinese and Indian antiquities, as well as sculpture, paintings, painted scrolls and screens, lacquerwork and Noh masks.

The Nezu Museum, Minami Aoyama Minato ku (open from 9-30 until 4, except Mondays), contains the famous "Iris Screen" by the painter Korin.

The Transportation Museum, Suda Cho, Chiyoda ku (open from 9-30 until 4, except Mondays) displays, amongst other things, the first Japanese railway engine to go into service, between Tokyo and Yokohama, in 1872.

The National Museum of Modern Art (open from 9 until 4-30) is devoted to twentieth-century Japanese art.

The Bridgestone Art Gallery, in the Ginza district (open from 10 until 5-30, except Mondays), contains a magnificent private collection of Western Art, including works by Rubens, Rembrandt, Guardi, the Impressionists, and Picasso—amongst others.

Lovers of the Japanese theatre will find much to interest them in the Theatre Museum, Totsuka machi (open from 9 until 4, except Mondays).

The Sumo Museum, Taito ku, Asakusa, Kuramae, traces the history of the sport.

TOKYO: City centre, main roads

TOKYO

Location and access: Island of Honshu, Kanto Region. Capital city of Japan, pop. 8,600,000 in the 23 inner districts or ku-area, 11,600,000 in the Tokyo Metropolitan Area. Haneda International Airport links Tokyo, via 26 international airlines, with all the major cities of the world. Internally, ANA (All Nippon Airways) provides daily connections to all main Japanese cities. Haneda Airport, 8 km south of the city, linked to it by monorail, trains, buses and taxis, will become the airport for international services, when the new Narita International Airport (60 km north of the city, in Chiba Prefecture) is finally opened. Visitors arriving by sea from America usually land at Yokahama, those from Europe and Australia, at Kobe. The "shinkansen" (bullet-train) station is 15 minutes by car from Kobe docks; the journey to Tokyo takes 3 ½ hours.

Distances from Tokyo: By road: Fukuoka, 1,150 km; Osaka 531; Kagoshima 1,463 km; Sapporo 1,110 km. By rail (shinkansen): Kyoto 513 km (2 hrs 50); Osaka 552 km (3 hrs 15); Hiroshima 895 km (5 hrs 10); Hakata 1,176 km (7 hrs); by other trains north; Sendai 352 km (4 hrs); Aomori 740 km (9 hrs).

Post code 100-180. Telephone code 03.

Travel agencies: JTB, 6-4 Marunouchi 1 chome; tel. 211-2701. Fujita Travel Service, Godo Building 2-10, Ginza 6 chome; tel. 572-0171.

Accommodation: Luxury hotels: *Imperial Hotel*****, 1-1, Uchisaiwai cho, Chiyoda ku; tel. 504-1111; 1,280 rooms; 5 min. by taxi from the station, 20 min. from the airport; a very high class establishment, a great rendezvous for foreigners in Tokyo; its restaurants are much esteemed by the Japanese too; its elegant rooms may be hired for receptions. *The New Otani***** (luxury), 4 Kioi cho, Chiyoda ku; tel. 265-1111; 2,050 rooms; 15 min. by taxi from the station and 30 min. from the airport; superbly situated; panoramic views from the revolving restaurant on 10th floor; its landscape garden makes it one of the most attractive hotels in Tokyo.

First category hotels: *Akasaka Tokyu Hotel****, 2-14-3, Nagata cho, Chiyoda ku; tel. 580-2311; 560 rooms; in city centre, 10 min. from station and 30 from airport. *Ginza Tokyu Hotel****, 5-15-9, Ginza, Chuo ku; tel. 541-2411; 450 rooms. *Keio Plaza Intercontinental Hotel****, 2-2-1 Nishi-Shinjuku; tel. 344-0111, 1,060 rooms; 5 min. walk from Shinjuku Station, 15 min. by taxi from the airport; one of the most modern—and tallest (45 floors)—hotels in Tokyo. *Okura Hotel****, 3, Aoi cho, Akasaka, Minato ku; tel. 582-0111; 930 rooms, gardens, swimming pools. *Pacific Hotel****, 3-13-3, Takanawa, Minato ku; tel. 445-6711, 950 rooms; 15 min. from the station by taxi, 15 min. from

BOOKS IN JAPANESE LIFE

■ *800 million books are published every year in Japan, a total sales value of some 500 billion yen. The whole nation seems to have become one of passionate readers. Students, employees and managers, teachers and sportsmen, they are all to be seen with their heads buried in a book as soon as they have a moment to spare. And all sorts of books: periodicals or history, classics, novels, scientific books, the latest best-sellers as well as masses of biographies. Magazines too have a growing readership. In 1976 1,500 titles appeared—a total of 1,250 million copies printed. There are 3,000 publishing houses, 12 of them very large indeed. Bookshops are always full of people browsing, reading, studying and buying. Why this thirst for knowledge in Japan? It stems largely from an ever-rising level of general education, from the demands of an information-orientated society, from a basic curiosity which is a fundamental trait of the Japanese character... Perhaps we should add too the feeling that extra knowledge will help an individual to rise more quickly to a higher level in his "work group".*

the airport. *Takanawa Prince Hotel*****, 3-13-1, Takanawa, Minato ku; tel. 445-5311; 458 rooms; 15 min. from the station and the airport; Japanese garden. *Tokyo Hilton Hotel*****, 2-10-3, Nagata cho, Chiyoda ku; tel. 581-4511; 470 rooms; 10 min. from the station by taxi, handy for the business quarter. *Tokyo Prince Hotel*****, 3-3-1, Shiba Park, Minato ku; tel. 434-4221; 500 rooms; close to Shiba Park, a typical first-class hotel, much frequented by Japanese for business meetings, weddings, etc; luxury shopping arcade.
Second category hotels: *Akasaka Prince Hotel****, 1, Kioi cho, Chiyoda ku; tel. 262-5151; 40 rooms; 10 min. from the station, 30 min. from the airport. *Azabu Prince Hotel****, 3-5-40, Minami-Azabu, Minato ku; tel. 473-1111; 30 rooms; quiet residential area. *Ginza Dai-Ichi Hotel****, 8-13-1, Ginza, Chuo-ku; tel. 542-5311; 800 rooms; in city centre near the Ginza and the business quarter. *Haneda Tokyu Hotel****, 2-8-6, Haneda kuko, Ota ku, Tokyo 144; tel. 747-0311; 290 rooms; opposite the airport. *Marunouchi Hotel****, 1-6-3, Marunouchi, Chiyoda ku; tel. 215-2151; 200 rooms; central, 2 min. walk from the station. *New Japan Hotel****, 2-13-8, Nagata cho, Chiyoda ku; tel. 581-5511; 470 rooms + 30 trad. style. *Palace Hotel****, 1-1-1, Marunouchi, Chiyoda ku; tel. 211-5211; 400 rooms; attractive view of the Imperial Palace Gardens; centrally situated. *Shiba Park Hotel****, 1-5-10, Shiba, Minato ku; tel. 433-4131; 300 rooms; near Tokyo Tower, good views of Shiba Park. *Shimbashi Dai-Ichi Hotel****, 1-2-6, Shimbashi, Minato ku; tel. 501-4411; 1,300 rooms; opposite Shimbashi Station; close to centre and business area. *Tokyo Fairmont Hotel****, 2-1-17, Kundan Minami, Chiyoda ku; tel. 262-1151; 240 rooms; attractive view of the Imperial Palace Gardens.
Third category hotels: *Diamond Hotel***, 25, Ichiban cho, Chiyoda ku; tel. 263-2211; 160 rooms; 10 min. from the station, 20 min. from the airport; Japanese, Western and Chinese cuisine. *Ginza Nikko Hotel***, 8-4-21, Ginza, Chuo ku; tel. 571-4911; 110 rooms; centrally situated, various restaurants, ideal businessmen's hotel. *Kokusai Kanko Hotel***, 1-8-3, Marunouchi, Chiyoda ku; tel. 215-3281; 95 rooms; opposite Tokyo Central Station and close to big stores on Nihombashi cho. *Takanawa Hotel***, 2-1-17, Takanawa, Minato ku; tel. 443-9251; 220 rooms; in the residential quarter, moderate charges, Japanese, Chinese and Western cuisine. *Takara Hotel***, 2-16-5, Higashi Ueno, Taito ku; tel. 831-1010; 90 rooms, 35 trad. style; close to Ueno Park and Station. *Tokyo Air Terminal Hotel***, 2-3-1, Haneda Kuko, Ota ku; tel. 747-01111; 50 rooms; in the Air Terminal Building (sound-proofed!), 20 min. from Tokyo by taxi. *Tokyo Hotel Urushima***, 2-5-23, Harumi, Chuo ku; tel.

JAPANESE SCRIPT

■ During the early centuries of the Christian era, Chinese script or Kanji, entered Japan. In this script the characters are figurative and there are as many characters as there are words. During the 7th century the Japanese, wishing to write their own language in a script better suited to its own particular character, adapted the Chinese system more phonetically, using a character to represent a syllable or a single vowel sound. Japanese script is known as Kana, sub-divided into Katakana and Hiragana; it has 100 syllables.
"Katakana" are often fragmentary characters, detached parts of Chinese ones. They are invariable and, in Japanese eyes, have the distinct disadvantage of not being calligraphic.
"Hiragana" characters were obtained by simplifying whole Chinese characters; they offer scope for elegant brush-strokes.
During the 10th and 11th centuries, the ladies of the Court wrote in Kana while their lords wrote in Kanji. Today every schoolchild in Japan learns the 100 Kana characters and some 1,900 of the 3,000 Kanji characters as well. Newspaper editors are not supposed to exceed these 1,900 basic characters. Sometimes Japanese have difficulty in reading a periodical because of the various ways in which it is possible to interpret a Chinese character in Japanese... The Japanese write from right to left and from top to bottom of the page.

533-31111; 1,000 rooms; close to the site of the Tokyo International Fair; views of bay and harbour; free bus service to Tokyo Station (20 min.). *Tokyo Kanko Hotel***, 4-8-10, Takanawa, Minato ku; tel. 443-1211; 150 rooms, 50 trad. style; quiet situation, opposite Shinagawa Station.

Fourth category hotels: *Daiei Hotel**, 1-15, Koishikawa, Bunkyo ku; tel. 813-6271; 90 rooms, 25 trad. style; in a quiet residential area near University and Koraku en Garden, 10 min. by car from Tokyo Station. *New Meguro Hotel**, 1-3-18, Chuo cho, Meguro ku, Tokyo 152; tel. 719-8121; 30 rooms; near Meguro Station. *San Bancho Hotel**, 1 Sanban cho, Chiyoda ku, Tokyo 102; tel. 262-333; 85 rooms; 7 min. by car from Tokyo Station. *Tokyo Hotel**, 2-7-8, Takanawa, Minato ku, Tokyo 108; tel. 447-5771; 46 rooms, 35 trad. style; the hotel looks like an old Japanese castle; 10 minutes by car from the station.

Restaurants: All the above hotels have restaurants serving Japanese, Chinese and Western food. There are several hundred foreign restaurants of all kinds in Tokyo. The following is a selection from the thousands of Japanese restaurants, classified according to their specialities:

"*Sukiyaki*": *Hasejin*, Minato ku, Azabudai 3-chome; tel. 582-7811. Imahan, Taito ku, Nishi-Asakusa 3-chome; tel. 841-1114. *Suehiro*, Chuo ku, Ginza 4-chome, Kintetsu Bldg.; tel. 562-0591. *Doh-Hana*, Bunkyo ku, 3-35-15 Yushima; tel. 831-5509.

"*Tempura*": *Inagiku*, Nihombashi Kayaba cho, 2, 6-chome; tel. 669-5501. *Kushi No Ippeieta*, 1-19-6, Shinjuku Bldg.; tel. 354-3400. *Ten-Ichi*, Ginza 6-chome, Miyuki/Namiki; tel. 571-1949-or Sony Bldg. and Toshiba Bldg. (Ginza). *Hashizen*, 1-7 Shimbashi, Minato ku; tel. 571-2700; speciality: "tempura" of shrimps. *Oijime*, 6-10 Kagurazaka, Shinjuku ku; tel. 260-2668; old restaurant dating from the Meiji period; specialities: "tempura" and "sushi". *Ten-Masa*, 3-38-5, Hongo, Bunkyo ku; tel. 811-0607; "tempura" Edo style, cooked in sesame oil.

"*Yakitori*" (grilled chicken): *Torikyu*, 3-9-11, Roppongi, Minato ku; tel. 402-4116. *Yakitori Nanbam*, 4-5-6, Roppongi, Minato ku; tel. 402-0606; specialities, skewered chicken and meat. *Torishige*, 2-8, Yoyogi, Shibuya ku; tel. 379-5188; working-class area, generous helpings. *Taiga*, 3-9-1, Akasaka, Kiyo Bldg., Minato ku; tel. 585-2934; yakitori and poultry dishes. *Totoya*, 6-21-1, Tsukiji, Chuo ku; tel. 541-8294; friendly atmosphere, near fish-market, generous helpings.

"*Robata-yaki*" (barbecue, Hokkaido style); *Inakaya*, 12-7, Akasaka 3-chome, Social Bldg.; tel. 586-3054; or, Rappongi 7-chome; tel. 405-9866. *Inaka*, Roof of Kokusai Bldg., Marunouchi; tel. 216-4606; branches in various districts.

Specialised restaurants: Iwashiya, 7-2 Ginza, Chuo ku; tel. 571-3000; sardines prepared in more than twenty different ways. *Nodaiwa*, 1-5, Higashi-Azabu, Minato ku; tel. 583-7852; Tokyo's leading eel restaurant. *Restaurants serving various cuisines:* Chin-

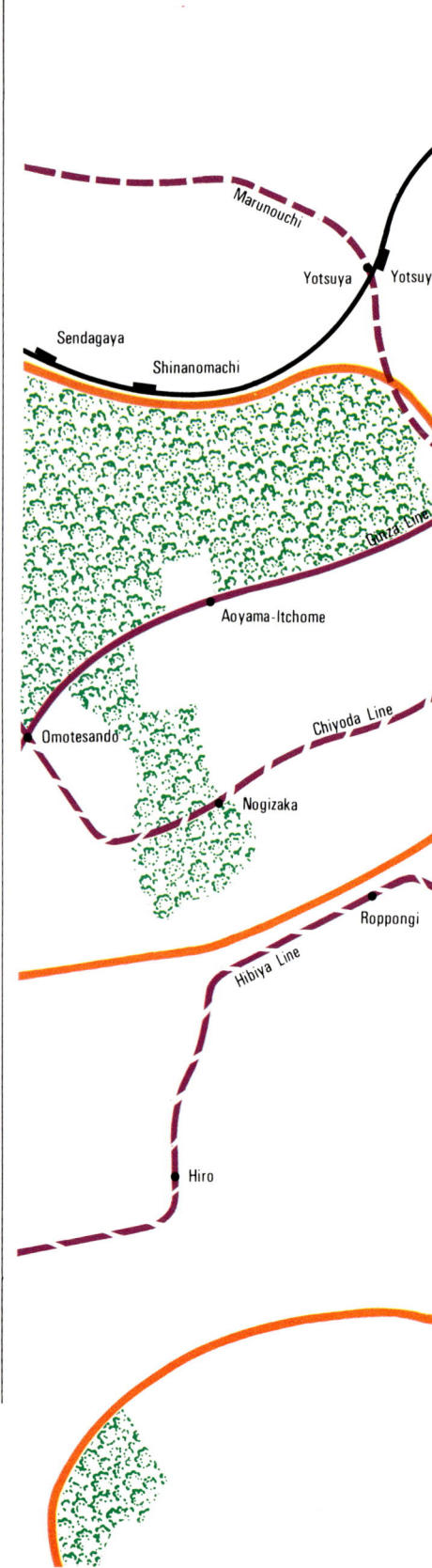

190 TOKYO

TOKYO 191

yokahama

zanso, 2-10-8, Sekiguchi, Bunkyo ku; tel 943-1111; pleasant atmosphere, Japanese garden, music, Japanese and Western specialities. *Furusato*, 3-4-1, Aobadai, Meguro ku, Shibuya; tel. 463-2310; Japanese specialities in an old-world Japanese setting (wooden house), music. *Tokyo Kaikan*, 3-2-1 Marunouchi, Chiyoda ku; tel. 215-2111; opposite Imperial Hotel; a variety of restaurants of different nationalities, under the same roof: Prunier of Paris, the Grill Rossini, Toh-en-—Chinese, Yachiyo—Japanese; international bar.

A few French restaurants in Tokyo: *Maxim's de Paris*, Ginza 3-chome, Sony Bldg.; luxury restaurant with chefs from Paris, evening dress expected, top prices. *L'Ecrin*, Ginza, Mikimoto Pearl Bldg.; tel. 561-9706; French chefs, very expensive. *Pente Rouge*, Akasaka, 3-5-4, Fuji Bldg.; tel. 586-5823; reasonable prices. *Rengaya*, Ginza 6-chome, Kajima Bldg., Namiki dori; tel. 573-0456. Run by Bocuse, cuisine lyonnaise, the best French wines in Tokyo. *Caput*, 1-4-1, Kasumigaseki, Chiyoda ku; tel. 591-4855; French cuisine in a Japanese atmosphere. *Bistro Lotus*, 6-7-8, Roppongi, JBP Bldg.; tel. 403-7666; intimate atmosphere.

A few reasonably-priced cabarets and night clubs: *Crown*, on the Ginza; variety shows at 9 and 10.30; tel. 572-5511. *Club Charon*, in the Akasaka district; intimate atmosphere, English-speaking hostesses; tel. 586-4480. *Club Casanova*, in the Roppongi district; very lively atmosphere, rock an' roll music; tel. 584-4558. *Mikado*, vast music-hall (one of the biggest in the world) in the Akasaka district, shows at 8 and 10; tel. 583-1101. *Golden Gessekai*, another gigantic music-hall in the Akasaka district.

Theatres: Ginza Noh Theatre, 6-5-15, Ginza, Chuo ku. Kabuki Theatre, 4-3, Ginza Higashi. National Bunraku Theatre, 13, Hayabusa cho.

Sport: Sumo, Kuramae Kokugikan, 2-1-9 (January, May, September).

Revue: Kokusai, 3-17-18, Nishi Asakusa, Taito ku.

Festivals: Sanja Matsuri, 17 and 18 May, at the Asakusa Shrine; Sanno Festival, 15 June, at the Hie Shrine.

■ Until 1850 Yokahama was a modest fishing village. Following the arrival of Commodore Perry and his ships (see: *Japan through the centuries*), under the terms of the Treaty of Kanagawa (26 July 1858) it was decided that the town of that name should be open to Westerners. In fact the Japanese were not at all happy about foreigners getting established in that port, so they made the excuse that space there was limited and the foreigners found themselves settled in the neighbouring marshy area of Yokahama where land was granted to them.

What seemed destined to fail became a success instead, for the area allocated to the foreigners turned out to be ideally suited for the construction of a deep-water port. The business area, near the wharves and known as the Bund, gradually expanded. The hill overlooking the harbour became the foreigners residential area; they built fine houses there and named it the "Bluff". On 5 September 1872 the Emperor himself paid a visit to Yokahama, to open Japan's first railway line, linking the city with the capital, Tokyo. This extraordinary occurence was the subject of much comment. One story relates how the dignitaries removed their shoes before boarding the train and were much surprised, at the end of their journey, not to find their "geta" waiting for them on the platform at the other end!

In 1889 the town became a municipality. After the earthquake of 1923 which was followed by disastrous fires (23,000 victims) the population of Yokahama had already reached 400,000. The city continued to spread and the harbour to grow. Incendiary raids in 1945 destroyed the greater part of the port area and the whole of the centre of the city. Today, Yokahama has 2,500,000 inhabitants; with its major industries, its maritime traffic and its university it is the third city of Japan.

Yokahama is not an easy city to visit, indeed it seems almost to have become a suburban extension of Tokyo—its port installations actually link up with those on Tokyo Bay.

There is still a slightly foreign air about Yokahama—about the "Bund" where there are many Western commercial houses, and about the "Bluff", Yamate Hill, the residential area. In the Western Cemetery, on the slopes of this hill, where 3,000 foreigners lie buried, the inscriptions on the gravestones, in many languages, recount better than any book the true history of Yokahama.

The most interesting part of Yokahama to explore on foot is the Chinese Quarter. Its narrow streets are lined with all sorts of little shops, bars and pachinko parlours. Many small restaurants, appreciably cheaper than those in Tokyo, offer the whole range of Chinese cooking—the highly-spiced dishes of Peking, the richer ones of Shanghai and the Cantonese cooking that Europeans tend to know best.

The great Sankei en Park (4 km from the city centre: bus from Sakuragicho Station) was laid out during the 19th century for a wealthy silk merchant, Hara Tomitaro. It is a garden full of valleys, woods and flowers and contains a collection of monuments brought from all over Japan.

There is a three-storeyed red pagoda of the Muromachi period (see: *Arts*), overlooking a pool covered with lotuses and waterlilies, and the Choshu kaku tea-house from the Nijo Castle in Kyoto. Near the Yanohara farmhouse, an eighteenth-century building from Hida province, stands the entrancing Yokobue tea-pavilion.

Some 400 metres south of the gardens, in a building designed by Adachi Kenzo in 1932, are the statues of the eight saints or sages of the world: Confucius, Socrates, Jesus Christ, and five Japanese: Prince Shotoku, and the priests Kobo Daishi, Shihran, Nichiren and Sakyamuni.

Finally, anyone interested in silk and its history should visit the Silk Museum, where a display of cocoons, tools and magnificent finished silks tells the whole story of the Japanese silk industry, from its origins to the present day.

YOKAHAMA

Location and access: Island of Honshu. Pop. 2,500,000; capital of Kanagawa Prefecture (pop. 6,300,000). A highly important industrial area and one of the leading ports of Japan. 24 km from Tokyo, 10 min. by train. Post code 213. Telephone code 045.

Travel agency: JTB, 75 Aioi cho 4 chome Naka ku; tel. 681-7541.

Accommodation: It is advisable to stay in Tokyo.

Festivals: Port Festival in mid-May; Black Ships' Festival, 14 July.

Souvenirs: Lacquerwork.

japanese journey

getting to japan

Where to get information:

Abroad: At branches of the Japan National Tourist Organisation (JNTO).
In England: 167 Regent Street, London W.1.; tel. 734-9638.
In U.S.A.: 45 Rockefeller Plaza, New York, N.Y. 10020.
333 North Michigan Avenue, Chicago, Ill., 60601.
In Australia: 115 Pitt Street, Sydney, New South Wales.
In Hong Kong: Peter Building, 58 Queen's Road, Hong Kong.
In France: 8, rue de Richelieu, Paris, 75001; tel. 742-20-19.
In Japan: in Tokyo, Kotani Bldg., 6-6 Yuraku cho, 1-chome, Chiyoda ku; tel. 502-1461/3 (open 9-12 and 1-5)
at Tokyo Airport (Haneda): Arrival Hall, Airport Terminal; tel. 747-0261 (open 9 until 8)
in Kyoto: Ground Floor, Kyoto Tower Bldg., Higashi Shiokoji cho, Shimogyo ku; tel. 371-5649 (open 9-12 and 1-5)

All JNTO offices in Japan issue, free, booklets containing practical information, lists of hotels, organised tours of various kinds, details of festivals, theatre programmes, information about hospitality in Japanese families. They do not sell travel or theatre tickets, nor do they make hotel reservations. They are closed on Saturday afternoons (except the branch at the airport), on Sundays and on national holidays.

Information can also be obtained from travel agencies, particularly from the Japan Travel Bureau (JTB), which has a branch in almost all Japanese cities. Their addresses are given with the practical information after entries for towns and sites.

Air services to Japan

Most leading international airlines have fairly frequent flights to Japan.
From Europe: Japan Air Lines, British Airways, Air France, Swissair and Alitalia all have at least three flights a week to Tokyo, thus there is a considerable number of flights daily and a great variety of routes available.
From America: Pan American and other leading airlines have daily flights to Japan.
From Australia: Quantas has two flights a week from Sydney to Tokyo and operates an association with Japan Air Lines on most other days.
Airline addresses in Tokyo:
Japan Air Lines: 7-3, Marunouchi 2-chome, Chiyoda-ku.
British Airways: Hibaya Park Building, 1-1-chome, Yuraku-cho.
Air France: Air France Building, 5-5, Akasaka 2-chome, Minato-ku.
Swissair: Hibiya Park Building, 1, Yuraku cho 1-chome, Chiyoda-ku.
Pan American: Kokusai Building, 1-1 Marunouchi 3-chome, Chiyoda-ku.
Quantas: Chamber of Commerce Building, 2-3 chome, Marunouchi, Chiyoda-ku.

Necessary documents

Passport: All foreigners entering Japan must have a valid passport, visaed at a Japanese Consulate. British passport holders do not need a visa for a stay of up to six months. Citizens of the United States, Australia, Canada, Hong Kong and Singapore all need visas, obtainable from their local Japanese Embassy or Consulate.

Vaccination certificate: All visitors must present an International Vaccination Certificate (yellow) on arrival, certifying that they have been vaccinated against smallpox (not more recently that 1 week, not less recently than 3 years). Cholera vaccination is also required of visitors coming from a contaminated area.

Customs regulations

Each traveller may import up to 3 bottles of spirits, 2 ounces of perfume, 400 cigarettes or 100 cigars, watches up to a value of 30,000 yen

*Previous page:
Fervent lovers of their country,
the Japanese take great pleasure
in exploring it at week-ends.
Here a group of town-dwellers revel in the contrasts
provided by the Arashiyama district of Kyoto.*

each, all free of duty. Personal effects and cameras for personal use are also admitted without charge.

When to go?

The pleasantest times to visit Japan are in the spring, when the cherries are in bloom, and in the autumn, when the colours are magnificent. June is the rainy month. The summer is very hot; the Japanese try to leave the coastal plains then and go up into the mountains. The island of Hokkaido is beautiful and interesting at the beginning of May, when the thaw is setting in on the lakes. Despite the bitter cold it is well worth visiting Hokkaido in January and February; the whole island is covered with snow, the northern ports are frozen up, and there are Snow Festivals in the cities.

Some useful addresses:

Embassies: Great-Britain, Ichiban-cho, Chiyoda-ku, Tokyo — *Germany,* 5-10 Minami Azabu 4-chome, Minato ku. - *Belgium* 5, Niban cho, Chiyoda ku. - *France,* 11-44, Minami Azabu 4-chome, Minato ku. - *Switzerland,* 9-12, Minami Azabu 3-chome, Minato ku. - *Canada:* 3-38 Akasaka 7-chome, Minato ku - tel. 408-21-01. - *United States:* 10-5 Akasaka 1-chome, Minato ku - tel. 583-71-41.

Foreign cultural institutes in Japan: *Maison franco-japonaise,* 3, Kanda Surugadai 2-chome Chiyoda ku. - *Institut Franco-japonais,* 15, Ichigaya Funagawara cho, Shinjuku ku. - *Goethe Institut,* 5-56, Akasaka 7-chome, Minato ku.

JAPANESE NATIONAL HOLIDAYS

1st January	New Year's Day.
15 January	Adult's Day; a day devoted to young people in their twenties.
11 February	Commemoration of the Foundation of Japan.
20 March	Spring Equinox.
29 April	The Emperor's Birthday.
3 May	Constitution Day.
5 May	Boys' Day.
15 September	Day of Respect for Old People.
23 or 24 September	Autumn Equinox.
10 October	Health and Sports Day.
3 November	Culture Day.
23 November	Labour Day.

If these holidays fall on a Sunday, then the following Monday is a holiday.

JAPAN AND THE FAR EAST

Japan

KURILES (U.S.S.R.)
Etorofu (Iturup)
Kunashiri (Kunashir)

HOKKAIDO

Sado

HONSHU
TOKYO
Oshima
Nijima
Miyake
Mikura
Hachijo

SHIKOKU

PACIFIC OCEAN

Japan and Europe
Latitude of Paris
FRANCE
SPAIN
Hondo
ITALY
Tokyo 137° 24' 15"
Kyūshū
AFRICA

Japan on the planisphere
Polar route
Paris
Moscow — Via Moscow
Via Peking — Peking
Via Bangkok — Bangkok
Anchorage
Vancouver
Toronto
Tokyo
San Francisco
Los Angeles
New York
Honolulu
Sydney

JAPANESE JOURNEY 199

planning the visit

Organised tours from overseas

Package tours organised by firms such as Jet Tours, Kuoni, and Air Tours usually last a fortnight and consist of a quick visit to Tokyo and to the essential sights in Kyoto, Nara, Nikko and Hiroshima.

Their advantage is that the journey is entirely trouble-free and there is no need to worry about any aspect of money or organisation. Such tours give the visitor number of stereotyped impressions; he may well conclude that Japan is a country of huge industrial cities in which a few islands of history and art miraculously survive, that the Japanese are a cold and distant people who shun contacts with foreigners, that they sleep on the floor in primitive fashion and eat a barbaric diet based on raw fish! Such visitors take all their meals in large Western-style hotels, with a night in a de-luxe "ryokan" to give them a little "local colour". They may attend a rather perfunctory tea-ceremony (organised specially for tourists) and a Noh or Kabuki play —which they will probably find peculiar, despite their English programme, and perhaps a show in a Tokyo night club, complete with geishas.

After such an initial glimpse the returning visitor may well have the impression that there is another Japan, behind all this, that it would be fascinating to explore and get to know, independently and at his own pace. The package tour will have been a useful preliminary however, and a second visit will be much easier thanks to the introduction it has provided.

Organised tours within Japan

There are many excellent tours available within Japan itself, organised by first class firms—Sunrise Holidays is one of the best. All hotel travel bureaux and travel agencies —notably JTB (Japan Travel Bureau)—are able to provide numerous prospectuses for one or two day tours, or longer round trips around Japan.

All such tours are most carefully prepared, the choice of hotels, restaurants, trains and coaches usually cannot be faulted. The visitor is courteously guided and thoroughly looked after. The programme is scrupulously adhered to and the tours normally go without a hitch of any kind. You may feel quite capable of exploring Tokyo and Kyoto for yourself, after all its very pleasant to stroll around and explore temples and gardens and museums on one's own. But if you are a relative novice to the country it is highly convenient to make a few longer trips knowing that you can leave all the arrangements to somebody else.

There are a few drawbacks however, namely:

Firstly: such tours do not yet cover the whole country—there are none, so far, to the island of Hokkaido or along the shores of the Sea of Japan.

Secondly: certain tours do neglect important sites. Thus the tour to shores of the Inland Sea takes in Okayama and Kurashiki but does not visit Himeji, the most splendid of all Japanese castles. Likewise, on the tour of Kyushu you will be taken to see the monkeys at Beppu (amusing enough) but not to see the marvellous stone Buddhas nearby. Also on Kyushu it's impossible to find one tour that includes both Kagoshima and Nagasaki—you have to decide which interests you the more and sacrifice the other.

Thirdly: such tours have the disadvantages or organised tours anywhere—the visitors remain together in an alien group and have little opportunity to sample the real life of the regions they are seeing.

Finally: Europeans may find the marvellous discipline of their Japanese fellow-travellers a little irksome at times. Smiling compliance is their rule; the slightest delay, say, in arriving back at the bus because you simply had to take that photograph, will be frowned on—after all the driver has work to do! (see: Good Advice and Good Manners)

Nevertheless as we said earlier, there are great advantages in taking one or two organised tours, so here is a selection of round trips offered by Sunrise Holidays, bookable through JTB offices, at the end of 1977. The prices quoted were those current at the end of 1977; they will serve as a guide at any rate. The letters correspond to those on the individual itineraries.

Tours starting from Tokyo:
A) Nikko: one day 12,700 yen (roughly £ 30 stg. or S45).
two days 18,000 yen.
B) Fuji and Hakone: one day 11,000 yen.
C) Hakone and Kamakura: one day.

Tours linking Tokyo and Kyoto:
C) 1st day: Kamakura and Hakone (one night).
D) 2nd day: Hakone and Kyoto—35,000 yen for two days.
E) linking with tours C) and D). 3rd day: Hakone-Toba-Ise-Kyoto: 40,000 yen.

Tours starting from Kyoto:
F) Nara: one day.
G) Iseshima: one day.

More extensive tours from Kyoto:
Along the shores of the Inland Sea:
H) 1st day: Kyoto, Osaka, Okayama, Kurashiki, Takamatsu.
2nd day: Takamatsu—boat trip to Osaka 40,000 yen.
I) 1st day: Osaka, Hiroshima, Miyajima.
2nd day: Hiroshima, Okayama, Kurashiki, Takamatsu.
3rd day: Takamatsu—boat trip to Osaka 60,000 yen.

Tours on the island of Kyushu:
J) 1st day: Kyoto, Osaka, Fukuoka (train).
2nd day: Fukuoka, Beppu (train).
3rd day: Beppu, Aso, Kumamoto (coach).
4th day: Kumamoto, Nagasaki.
5th day: Nagasaki, Fukuoka (train). Cost, from Kyoto or Osaka, 127,000 yen.
K) This tour is a continuation of the preceding one:
5th day: arrive at Okayama in the evening. 6th day: Okayama, Kurashiki, Takamatsu. 7th day: Takamatsu—boat trip to Osaka. Cost: for the seven days 163,000 yen.
L) To Kyushu by jet:
1st day: Kyoto, Osaka, Fukuoka by jet. 2nd day: Fukuoka, Beppu. On the following days this tour links up with the preceding tours J) and K). The return journeys are made by train. Total cost: 160,000 yen.

Travelling independently

An independent visit to Japan is fascinating and perfectly feasible on certain conditions. The visitor will discover that travel agency staff, shop assistants in big stores in the larger cities, as well as those in souvenir shops, radio shops and jewellers', all speak a certain amount of English. Staff at large and medium hotel reception desks, as well as those in the larger "ryokan" generally speak English well. There is not much problem in large restaurants in big cities either. However, once you leave the main tourist areas you will meet comparatively few English speakers; but the Japanese are extremely helpful and a few gestures and smiles will usually enable you to make yourself understood.

Unless you read and speak some Japanese or have a Japanese friend to guide you, it is advisable to give up any idea of driving in Japan. All traffic signs and signs indicating the names of towns are always written in Japanese alone. Even if you have an excellent map it is virtually impossible to avoid making mistakes and finding yourself back where you started after an hour's fruitless driving round and round. The Japanese railway system is excellent but the roads are less good; traffic often moves very slowly and traffic jams are frequent. Visitors are strongly recommended to use planes and trains, and buses for short distances.

It is important to realise that once you leave Tokyo you must rely on yourself to find your way. The ordinary people in Japan are often very kind, but there are many occasions when gestures are just not enough

POLITICAL JAPAN: provinces, cities, railways

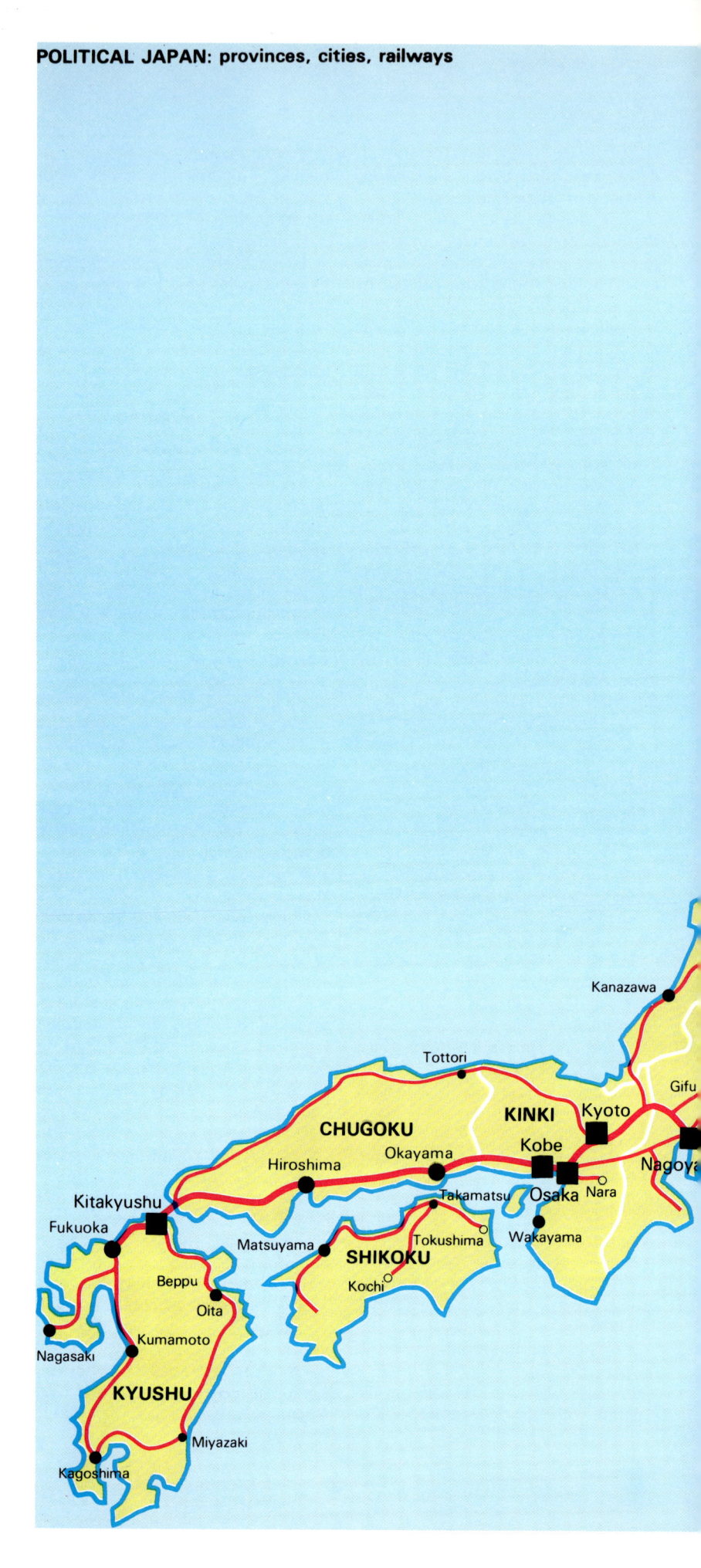

202 JAPANESE JOURNEY

Japan

Cities and Regions

Hokkaido
- Abashiri
- Asahikawa
- Sapporo
- Kushiro
- Noboribetsu
- Hakodate

Tohoku
- Aomori
- Akita
- Sendai

Hokuriku
- Niigata

Kanto
- Nikko
- Utsunomiya
- Nagano
- Urawa
- Tokyo
- Chiba
- Yokohama
- Hakone

Chubu
- Takayama
- Shizuoka

Legend

Railways
- Shinkansen
- Other lines

Provinces

Population of principal cities (capitals of prefectures)
- ■ Cities of more than 5 million inhabitants
- ■ Cities of 1 to 5 million inhabitants
- ● Cities of 500,000 to 1 million inhabitants
- • Cities of 300,000 to 500,000 inhabitants
- · Cities of less than 300,000 inhabitants

Scale 1 : 6 000 000
0   50   100   150 km

JAPANESE JOURNEY

—as the following story shows. A visitor to Kyoto wanted to cross through the station rather than take the longer way round following the streets outside. He looked in vain for a pedestrian subway, climbed a gangway instead, and found himself lost in a shopping area. He sought advice at three ticket windows, but the officials spoke only Japanese. He went up to a man at the barrier and attempted to explain his problem. The man smiled and said "Hai" (yes), then pointed to a row of machines selling tickets, probably for all the suburbs of Kyoto? Suddenly struck by a flash of inspiration, the visitor realised that he had to buy a ticket in order to cross the station. But from which machine?—since everything was written in Japanese it was impossible to know. After half an hour he was just about giving up when a Japanese approached him and, in halting English, managed to put him on the right track.

If crossing Kyoto Station is fraught with such problems, it's difficult to believe that a journey to Hokkaido would be without its hazards—unless it was very carefully prepared beforehand.

The visitor should, in fact, prepare his journey carefully and then check his plans with a travel agency in one of the larger cities. JTB offices are highly recommended. They exist practically everywhere and their staff speak English. The visitor will have to spend quite a while explaining to the person in the travel bureau exactly what he wants to see, where he plans to spend the nights, whether he wants to leave in the morning or the evening, and so forth. The travel bureau will then need an hour or two to make the arrangements. When the visitor returns for his tickets he will be handed a great wad of hotel reservations, as well as restaurant and train tickets—all written entirely in Japanese. It is essential that to note down there and then the times of trains, the names of the hotels and whether the meals are included in the price; these details may seem irrelevant at the beginning, but the success of the holiday may well depend on getting them right. The travel agency will give you a map, in Japanese, of every town you plan to visit. You should get them to mark, in Japanese, the name and address of your hotel as well as any points of interest you may wish to see. Any taxi driver will then be able to take you exactly where you want to go. Such preparations are an essential part of a successful journey. The travel agencies of course wish to sell as many of their conducted tours as possible, so they will try to discourage you from travelling independently (frequently exaggerating the risks involved); don't be put off, but remember: you have to insist!

Japan by train

The Japanese railway network is highly-developed. The Japanese themselves use it a lot and the trains are often full. It is essential to buy one's ticket and book one's seat several days in advance, for express services. If you are touring a particular area you will get a "basic" ticket covering the whole journey—it must never be given up, nor lost—as well as separate tickets for each stage of the journey, which also serve to reserve seats on the train. Japanese trains are clean and comfortable. Cleaning women pass through frequently, to pick up papers and empty bottles. The "shinkansen" is the fastest train in the world (see special entry). All trains have eating and drinking facilities (see: *Food*). Before the train arrives at a station there is a burst of music from the loudspeaker and then a lengthy speech in Japanese. This consists of a list of possible connections from the next station, various polite phrases and injunctions to passengers not to leave anything behind when they leave the train. In almost all stations the name of the town is written up in "romaji" (Latin lettering), which makes it easier for the foreigner to identify. Express trains stop exactly opposite numbers written on the platform which correspond to the numbers of the seats booked by passengers. This avoids a scramble to board the train. The same cannot be said of the local trains, crowded with schoolchildren and workers during the rush hours, all pressing forward in their anxiety to get on board. Large suitcases are highly inadvisable. Porters are almost non-existent, except in Tokyo. Few stations have lifts or escalators, there are many gangways to negotiate as well as steep steps, and the trains, even the modern "shinkansen", are not built to take much luggage.

If you have a lot of luggage then it is advisable to choose seats at the back of the coach, where there is often a space to stow suitcases. The Japanese tend to take lots of small parcels with them when they travel, which present few problems. Visitors are advised to take as little as possible with them to Japan and, once there, to buy a big bag on wheels. They can then leave their large suitcase in their hotel in Tokyo and travel all over Japan using their marvellous bag on wheels, much less tiring!

Japan by plane

Japan has excellent internal air-services. The main lines are:
JAL (Japan Air Lines), with flights from Tokyo to Sapporo, Osaka, Fukuoka and Okinawa.
ANA (All Nippon Airways)
TDA (Toa Domestic Airlines)
It is advisable to book seats a few days in advance.

Taxis in Japan

No country in the world has as many taxis as Japan. You never have to walk as far as a taxi rank, cabs can always easily be hailed from the street. Beware! The driver controls the opening of the rear doors; many foreigners have suffered unexpected injuries because of this—so watch out! It's normally up to passengers to stow their own luggage in the boot of

THE SHINKANSEN

■ The "shinkansen", the fastest train in the world, has a maximum speed of 210 kilometres per hour, and covers the distance between Tokyo and Hakata (1,176 km) in 7 hours, serving 28 cities in the Japanese industrial zone. There are two types of shinkansen: the "hikari" and the "kodama". The "hikari" stops only at the most important places the "kodama" stops everywhere. Every ten minutes, between 6 in the morning and 5 in the evening, a train leaves each terminus. The trains are identical, each consisting of 16 coaches, air-conditioned and pressurised. Both 1st and 2nd class coaches are fitted with swivel seats. The "hikari" has a dining car, serving Japanese and Western food, and a buffet car. The "kodama" has no dining cars but is provided with stands stocked with wrapped food of various kinds. Both kinds of trains have a trolley service for the sale of drinks, souvenirs and newspapers. Speedometers enabling passengers to check the speed of the train and radio-telephones, connecting with the cities served, are to be found in the buffet car. The trains run smoothly on welded rails. There are numerous safety devices. In Tokyo the position of every train is indicated on an illuminated chart. An automatic system, connecting the signals along the route with the trains' braking mechanism, enables the train to be slowed down if it should exceed its correct speed. Along the line there are seismographs, anemometers and rain gauges. In the event of a severe earthquake, the electric current is immediately turned off. Seats in the leading coaches are available on a come-as-you-please basis; the others are bookable 7 days ahead, through a computer system operated at the stations and JTB offices.

NATURAL FEATURES

SIBERIA

Vladivostok

MANCHURIA

SEA OF JAPAN (NIHON KAI)

KOREA

Fusan

Okinoshima Islands

HONSHU

Tsushima Islands

Hiroshima Okayama Kobe Nagoya
                          Kyoto
                          Osaka
                          Ise-Shima N.P.

Kitakyushu

Inland Sea Maritime National Park

Aso N.P.

Nagasaki

Mount Aso

SHIKOKU

Kirishima

Kagoshima Kirishima-Yaku N.P.
Sakurajima

KYUSHU

206 JAPANESE JOURNEY

HOKKAIDO

- Daisetsuzan N.P.
- Akan National Park
- Sapporo
- Showa Shinzan
- Shikotsu-Toya N.P.
- Koma Ga Take
- Sendai
- Nikko N.P.
- Asama
- Fuji Yama
- TOKYO
- Yokohama
- Fuji-Hakone-Izo N.P.
- Mijara (I. Oshima)

PACIFIC OCEAN
(TAI HEI YO)

Legend:
- Plains
- ▲ Volcano
- ○ Mineral spring
- ✳ Major National Park

1 : 6 000 000
0   50   100   150 km

JAPANESE JOURNEY 207

the car, the white gloved drivers rarely perform this service. They also often do not speak English, so it's essential to have the names of streets, temples and so on, written down in Japanese—also your hotel card for the return journey. Hotel receptionist-interpreters will frequently help out with instructions to taxi drivers.

Tour coaches

There are thousands of organised trips every week-day and at weekends—for the Japanese are themselves avid sightseers. There are also regular buses serving the main beauty spots in Kyushu and Hokkaido. The vehicles generally travel slowly (40 km per hour) and make frequent stops to allow passengers to get out and enjoy the view. They usually stop for a lunch break (see: *Food*).

There are always two staff aboard; the driver, the hostess—white gloved and uniformed—and sometimes an "under hostess" who keeps the coach clean and wipes the windows whenever it stops.

Often the hostess will give a running commentary through a microphone—with burst of appropriate poetry and song. She will often keep it up for hours, even though most passengers may be talking or asleep. To a foreigner it will perhaps seem charming at first, though his enthusiasm may pale as the torrent of Japanese continues... but still it's all part of the local scene. Other Japanese passengers are usually most welcoming to foreign visitors.

Six suggested itineraries

Here are six suggested Japanese itineraries, with accompanying maps.

If the visitor followed them all it would take six weeks; they can of course be abbreviated to fit the time available. They are quite feasible trips and they are described in fair detail, so you can adapt them as you wish. The prices quoted were those current at the end of 1977; they will give some idea of relative costs.

1. A week based on Tokyo: a week's stay in the capital, with some or all of the following tours. Nikko—full day (see organised tour); Kamakura-Hakone—full day (see organised tour); Mount Fuji—full day (see organised tour); Yokahama—an independent day-tour. Leave Tokyo from Shinagawa Station—train journey of about ½ hour to Yokahama—visit the city—lunch in the Chinese quarter—return to Tokyo for the evening.

2. Nine days on the island of Hokkaido (Tour No 1 on the map):

First day: Tokyo-Sendai. Leave Tokyo from Ueno Station, journey takes about 4 hours (inclusive ticket Tokyo-Chitose 7,300 yen; Tokyo-Sendai 1,600 yen)—taxi from Sendai Station to hotel (2 nights approx. 18,000 for two)—tour of city by taxi during afternoon (ask hotel reception to arrange the charge with the driver and explain what you wish to see).

Second day: Organised tour to Matsushima (coach leaves from near the station early in the morning, cost 2,250 yen). A good chance to join a group of happy Japanese holiday-makers. The coach will stop for lunch at a good view-point; cost of lunch ("soba" and "sushi") not included, but boat trip around the islands is. Return to Sendai early evening.

Third day: Sendai-Hakodate. Leave Sendai by train "Hatsukari No !"—lunch on train (see: *Food*) —arrive at Aomori (train Sendai-Aomori 1,600 yen)—take gangway to ferry—arrive Hakodate late evening —hotel ("Royal Hakodate"—8,500 yen for two)—dinner in small bistro nearby.

4th day: Hakodate-Noboribetsu. Visit Hakodate in morning—train to Noboribetsu early afternoon, ("Hokuto No 2", 3 hrs, cost 1,200 yen) —sit on right-hand side of carriage for best views of coast—arrive and transfer to hotel ("Grand Hotel Ryokan"—dinner, room and breakfast for two 36,000 yen)—settle into room, bathe in hot spring, dine (see: *Accommodation* if staying in ryokan).

208 JAPANESE JOURNEY

5th day: Noboribetsu, tour to Shiraoi. Visit Noboribetsu in morning—climb Shirorei—afternoon coach tour to Shiraoi, Ainu village (see entries)—ask about buses to village (taxis are expensive—9,000 yen for 2 hrs).

6th day: Noboribetsu-Akan. Early train to Chitose (Sapporo airport)—taxi to airport itself—early lunch—fly to Kushiro (flight 8,000 yen)—leave by bus for Akan—hotel ("Newakan", ryokan; dinner, room and breakfast, 20,000 yen for two)—perhaps in time for the last boat-trip on the lake (4 o'clock).

7th day: Akan-Abashiri. Early departure by coach (tickets at the station) for Bihoro (6hrs, 3,000 yen)—lunch off snacks bought at stops en route—train straight on to Abashiri (½ hr)—hotel ("View Park", ryokan; dinner, room and breakfast, 18,000 yen for two)—hire taxi through hotel reception for tour of Abashiri.

8th day: Abashiri-Sapporo. Early train to Sapporo (6 hrs, 4,000 yen)—lunch off snacks on train—hotel ("Park Hotel", 12,000 yen for two; "Sapporo Zeniku Hotel", 10,000 yen for two)—visit city—dinner, walk around Sapporo by night.

9 th day: Sapporo-Osaka-Kyoto. Early morning visit to Moiwa-Yama—leave by plane from Chitose airport early afternoon (buses collect airport passengers at hotels at stated times)—arrive Osaka (2 hrs; 28,000 yen)—bus to Kyoto.

3. *A week in Kyoto.*

At least three days must be allowed to see the city (see suggested visits in entry for Kyoto). Some or all of the following tours are recommended: Nara—full day (see organised tour); Ise Shima—full day (see organised tour); Osaka—full day, by train; Koya San—full day; take an early morning train to Osaka—taxi at Osaka to Namba Station—an hour's journey to Gokurakubashi—take funicular to Koya San. It's a good

A RECIPE FOR SUKIYAKI

■ *This dish is prepared at table in front of one's guests. For four people you will need:*

½ kilo of fillet of beef. Ask your butcher to put the meat into his freezer for a while and then cut it into very thin slices.

1 packet of transparent noodles, "shirataki". Cook them quickly in boiling water and drain.

1 box of bamboo shoots, sliced.

1 large onion, finely sliced.

1 packet of Chinese mushrooms, soaked and drained (or a quantity of fresh mushrooms, sliced).

2 boxes of "tofu" (soya bean) paste, diced.

Some chopped leek or chives, and a few cabbage or chrysanthemum leaves.

Beef fat. 1 egg per person. Soy sauce, sugar, sake.

1. Arrange the raw meat and the vegetables on a dish.

2. Ask each guest to break an egg into his bowl and beat it up with his chopsticks.

3. Melt the beef dripping in a pan, over a burner, on the table. Add 5 slices of meat, 5 dessert spoons of soy sauce and 3 dessert spoons of sugar. Let them cook for a minute, turning the meat over with chopsticks. Add some vegetables and noodles, together with 4 spoons of sake. Turn the vegetables in the sauce. Cook for 5 minutes. Serve. The guests dip their meat into their bowls of beaten egg and then eat it. The rest of the ingredients are prepared in the same way; there should be enough for three servings. Hot sake should be drunk with Sukiyaki.

CENTRES OF INTEREST AND IMPORTANT MONUMENTS

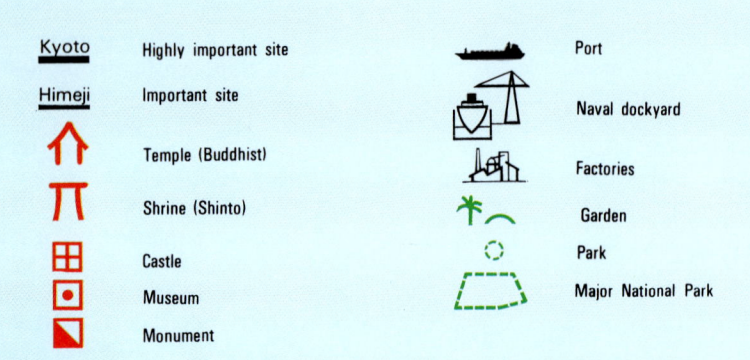

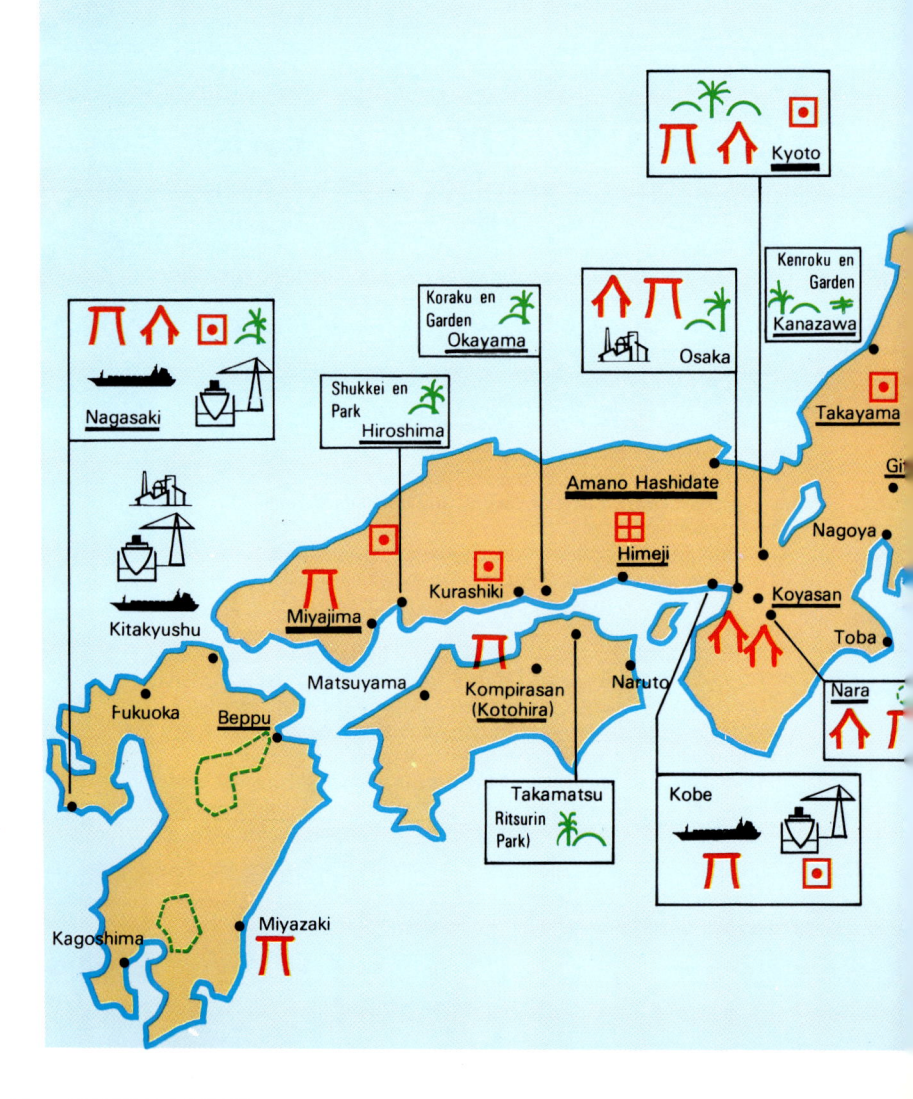

Kyoto	Highly important site
Himeji	Important site
⚟	Temple (Buddhist)
⛩	Shrine (Shinto)
⊞	Castle
⊡	Museum
▮	Monument
🚢	Port
⚓	Naval dockyard
🏭	Factories
🌴	Garden
○	Park
⬠	Major National Park

210 JAPANESE JOURNEY

JAPANESE JOURNEY 211

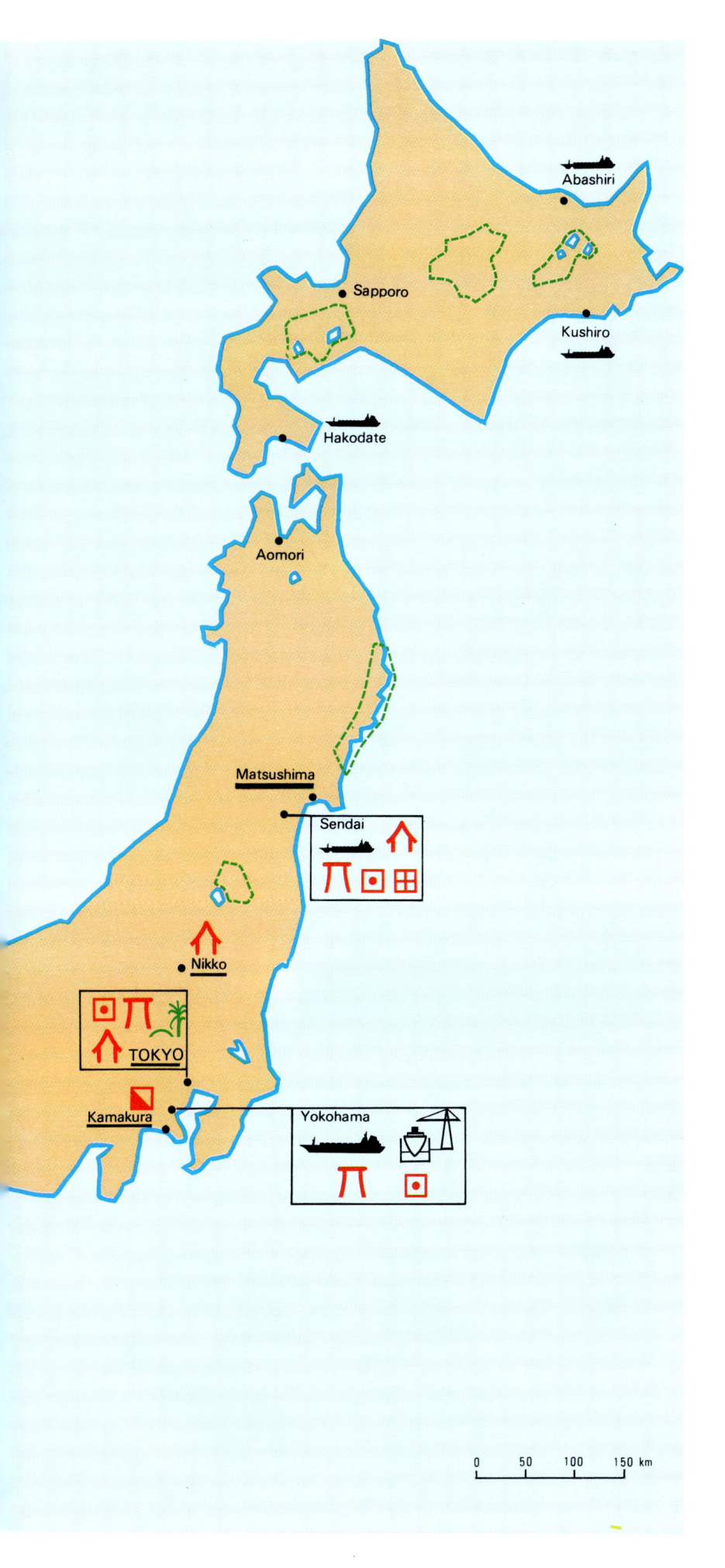

idea to enquire beforehand at the JTB office in Kyoto or Osaka, as there are sometimes bus trips to Koya San from Osaka itself.

4. *Seven days on the shores of the Inland Sea,* starting from Kyoto (Tour No 2 on the map):
An inclusive ticket Kyoto-Hiroshima costs 3,800 yen.
1st day: Kyoto-Kobe. Early train from Kyoto to Kobe (1 hr)—visit Kobe (see entry) —climb to the Rokka San in late afternoon—hotel ("New Port Hotel", 9,000 yen for two; "Rokka San Hotel", 8,000 yen).
2nd day: Kobe-Himeji-Kurashiki. Early train from Kobe to Himeji (½ hr)—visit Himeji Castle—train to Kurashiki (50 mins)—visit Kurashiki —hotel ("Kurashiki Kokusai", 9,000 yen for two).
3rd day: Kurashiki-Okayama-Takamatsu-Naruto. Before making the trip to Naruto enquire at the JTB office about the tides. Leave Kurashiki by early train for Okayama (15 mins)—visit castle and gardens —lunch in a snack bar near gardens —mid-afternoon train to Uno—transfer to hovercraft (a simple matter, it's at the end of the platform)— hovercraft to Takamatsu (1,000 yen)—bus to Naruto (Takushima line, bus stop opposite station, in front of "Grand Hotel"; buy tickets on bus)—arrive early evening.—taxi to hotel ("Shishishuen" a simple ryokan, recommended for its position overlooking whirlpools; dinner, room and breakfast, 12,000 yen for two; the meal is fairly Spartan: have some biscuits with you!).
4th day: Naruto-Takamatsu. An early start to see whirlpools from view-point near hotel —taxi to landing stage for boats to Fukura, on Away Shima Island (no problem with taxi drivers here, they're used to this trip)—boat trip across, via whirlpools—bus to Takamatsu—arrive hotel ("Grand Hotel", 16,000 yen for room for two nights for two)—af-

THE AINU BEAR FESTIVAL

■ *The preliminary to this festival is a thoroughly good wash; something that the Ainu do not indulge in frequently... A bear has been reared specially for the occasion, to serve both as god and victim. Since the cub was taken from its mother very young, an Ainu woman has been suckling it. The ceremony begins with the men making a sacrifice to the god of fire. Then the bear is dragged from its cage, a piece of wood is thrust into its mouth, it is thrown over onto its back, its neck against a log, with nine men kneeling on it, stifling it. This cruel death is inflicted entirely without malice on the part of the killers. The Ainu are making an offering of this one bear to the whole bear family in order to expiate all the other deaths they have caused, as a means of pardon and reconciliation with the whole species. While the men are busy suffocating the bear, the women and girls perform a dance around them, weeping, striking them as they pass. The following day the bear is cut into pieces. Its blood is drunk with great relish. Its liver is eaten, raw, with salt, its brains are drunk with sake. The man who cut up the animal is given the eye eat as a reward. Thus all who participate become imbued with the flesh and nature of the animal—become bears themselves so to speak. Meanwhile the women have continued their dancing, almost unceasingly, singing in a monotonous tone all the while. This is the only great celebration in the whole year amongst these pathetic people.*
G. BOUSQUET,
Le Japon de nos Jours, 1870

ternoon visit to Ritsurin Park—late afternoon to Yashima Plateau to see the sunset.

5th day: Takamatsu, excursion to Koto-Hira. Trains leave every half hour, station opposite "Grand Hotel", 1 hr journey—in Koto-Hira visit Kompira San—lunch in local inn—return to Takamatsu for night.

6th day: Takamatsu-Matsuyama-Hiroshima. Leave on very early train for Matsuyama (3 hrs) — bus to Kanko pier (50 mins) — hovercraft to Hiroshima (1 hr; 3,400 yen)— arrive hotel ("Hiroshima Grand Hotel", 9,000 yen for two)—visit Hiroshima.

7th day: Miyajima Island and Itsukushima Shrine. Leave by an early train for Miyakjimogushi (½ hr)—ferry to Miyajima (every ten mins)—day on the island (see entry) returning in the evening. (It is possible to vary this: a. by taking the "shinkansen" back to Osaka and Kyoto; N.B. the last through train leaves Hiroshima at 7.12, arriving at Osaka at 9.10 and Kyoto at 9.29. b. by continuing your journey with Itinerary No 5, Tour No 3 on the map.

5. Island of Kyushu, returning along the shores of the Sea of Japan as far as Amano Hashidate and Kyoto (Tour No 3 on the map):

An inclusive ticket costs 7,000 yen.

1st day: Beppu and the Buddhas of Usuki. Arrive at Beppu by air and train (3 hrs) from Tokyo or Osaka-Fukuoka, or, better, by air from Osaka-Oita (1 hr)—or continue on from Itinerary No 4 (Tour No 2 on the map). Take the night boat from Hiroshima to Beppu (it normally leaves at 9 pm, you have to be on board by 8; 14,800 yen in a cabin for two, less if you sleep on mats in a dormitory)—arrive at Beppu at 5 in the morning—taxi to hotel ("Suginoi", 13,000 yen for two for two nights)—hotel room probably not free until early afternoon, so leave luggage and take a bathe in a hot spring, have breakfast and go to see the

BIRTH OF A PEARL

■ *Cultured pearls are produced by artificial stimulation of a natural process. Into a "mother oyster", three years old, is placed a tiny round pearl made from the shell of a Mississippi mollusc, together with a little piece of the veil of another oyster. The oyster reacts by secreting nacre in order to cover this irritating foreign body. The pearl oysters, kept in cages, are attached to wooden rafts that are then moored in sheltered bays chosen for their gentle currents. Several times a year the cages are cleaned of the seaweed that has begun to grow on them. After three years the pearls themselves are extracted, cleaned, graded and polished. Losses are enormous. From a million treated oysters only about ten thousand saleable pearls are produced. Some seven thousand women are employed as divers, diving from three to four hours a day to harvest "wild" oysters; it takes about ten years to become a good diver. The oyster beds are faced with the threat of typhoons every year. In 1974, 47 tons of pearls were produced, for export to Western Germany, Switzerland, Hong-Kong and the United States. Apart from Ise Shima, the important Japanese pearl-producing districts are Kumamoto and Nagasaki.*

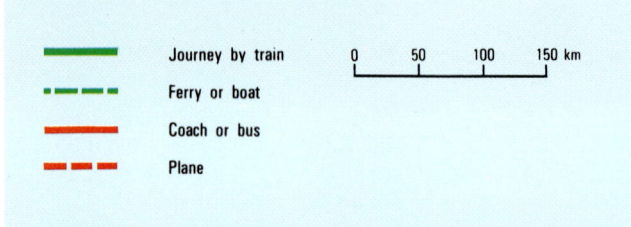

JAPAN IN 40 DAYS

— Journey by train
-- Ferry or boat
— Coach or bus
-- Plane

0 50 100 150 km

A 4 day round trip through the Alps

8 days in Kyushu, returning via the Sea of Japan

A week in Kyoto

Round Trip II
A week on the shores of the Inland Sea

Kanazawa
Amano Hashidate
Himeji
Kurashiki
Kobe
Kyoto
Gifu
Hiroshima
Okayama
Osaka
Miyajima
Takamatsu
Nara
Toba
Kotohira
Naruto
Kitakyushu
Koyasan
Fukuoka
Beppu Matsuyama
SHIKOKU
Nagasaki Kumamoto
KYUSHU
Kagoshima Miyazaki

214 JAPANESE JOURNEY

JAPANESE JOURNEY 215

"Hells" (see entry for Beppu)—afternoon train trip to Usuki to see Buddhas—return to Beppu for night.

2nd day: Beppu, trip to Mount Aso. Whole day coach tour to Mount Aso (coach leaves from opposite the station)—usual jolly Japanese company; lunch of "soba" or soup (see: *Food*)—return to Beppu for night.

3rd day: Beppu-Miyazaki: Leave Beppu by early train for Miyazaki (4 hrs)—hotel ("Phoenix", 8,500 yen for two)—taxi trip along the southern coast (ask hotel reception to help you fix price with driver)—visit Miyazaki.

4th day: Miyazaki-Kagoshima. Leave Miyazaki by early train for Kagoshima (2 hr 15 min)—hotel ("Kagoshima Hayashida", 6,000 yen for two)—take the trip to the volcano.

5th day: Kagoshima-Nagasaki. Early morning train from Nishi Station, change trains at Kumamoto for Misumi (Kagoshima-Misumi takes 4 hr 15 min)—ferry to Shimbara—bus from Shimbara (opposite station) to Nagasaki (3 hrs)—hotel ("Grand Hotel", 18,000 yen for two for two nights).

6th day: Visit Nagasaki.

7th day: Nagasaki-Fukuoka. Leave by early train (2½ hrs) for Fukuoka—hotel ("Station Plaza", 9,000 yen for two)—visit city.

8th day: Fukuoka-Amano Hashidate. Early morning train ("matsukaze") to Toyooka—packed lunch on train—(10 hrs)—dinner at Toyooka in restaurant opposite the station—train to Amano Hashidate (1½ hrs)—hotel ("Gemmyodan", an excellent ryokan, 9,000 yen for two).

9th day: Amano Hashidate.—return to Kyoto. Visit city in the morning—train to Kyoto late afternoon (2½ hrs).

6. *Kyoto-Gifu-Takayama-Kanazawa-Kyoto* (Tour No 4 on the map):

1st day: Kyoto-Gifu (with possible continuation to Takayama). Early morning train from Kyoto to Gifu (1½ hrs)—hotel ("Grand Hotel", 9,000 yen for two)—visit Gifu—evening trip to watch cormorant fishing (from 15 May until 15 October)—or afternoon train to Takayama (2½ hrs).

2nd day: Gifu-Takayama. Morning train from Gifu to Takayama (2 hrs 45 mins)—hotel ("Hida", 7,000 yen for two)—afternoon, visit city.

3rd day: Takayama-Kanazawa. Day in Takayama—early evening train to Kanazawa (2½ hrs)—hotel ("Kanazawa Sky Hotel", 8,000 yen for two).

4th day: Kanazawa-Kyoto. Morning in Kanazawa—afternoon train to Kyoto (2½ hrs, approx.). □

everyday life in japan

Money

The exchange-rate of the yen is constantly changing. Early in 1978 it was running at around 400 yen to the £ sterling.
Tipping: This is not a Japanese habit. A service charge of 10 to 15 per cent is added in luxury shops and restaurants, as well as a 10 per cent tax on all food and drink costing more than 1,500 yen.

Times and calendar

Time: Japanese time is usually 9 hours ahead of British, 8 hours ahead of most of Western Europe.
Banks: Normally open from 9 until 3, except Saturdays, 9-12 noon. Closed on Sundays and public holidays. Airport banks are usually open 24 hours a day.
Offices: Normally open from 8.30 until 5. Closed on Saturdays, Sundays and public holidays.
Shops: Normally open from 10 until 8 in the evening.
Big Stores: Usually open from 10 or 10-30 until 6 or 6-30. Open on Sundays and public holidays. They close on one day during the week.
Calendar: 1978 is the 53rd year of the reign of Emperor Hirohito or Showa. The rubber stamps of which the Japanese are so fond—they collect them enthusiastically at tourist sights—all bear the number 53. Motor vehicles too are all registered "53".

Public Services

Electricity: 110 volts is standard throughout Japan; they use American-style plugs.
Postal services: Post Offices are open from 8 until 8 in the evening, except Saturday afternoons, Sundays

THE TRADITIONAL FESTIVALS

January: 1-3 exchange, of New Year preetings; visits to shrines and temples.
February: At the beginning of the month, Sapporo Snow Festival.
March: 3, "Hina-matsuri" (Doll Festival), when little girls display their dolls.
April: 14-15, "Takayama Matsuri", the Hie Shrine Festival, at Takayama.
May: 15, "Aoi Matsuri", Hollyhock Festival, in Kyoto, splendid processions. 17-18, "Sanja Matsuri" at the Asakusa Shrine, portable altars are carried through the streets. 18, Nikko Spring Festival, procession of 1,000 samurai.
June: 14, Rice Planting Festival, at the Sumiyoshi Shrine, in Osaka. 15, Sanno Festival at the Hie Shrine in Tokyo; procession of portable altars.
July: 13-15, "Bon" Festival (see: Religion). 16-17, "Gion Matsuri", in Kyoto, procession with floats.
August: 6-8, "Tanabata" or "Star Festival", at Sendai.
September: 16, Festival at the Tsurugaoka Hachimangu Shrine, at Kamakura, demonstration of archery on horseback.
October: 7-9, "Okunchi" or Suwa Shrine Festival, in Nagasaki. 9-10, "Takayama Matsuri" or Hachiman Shrine Festival at Takayama. 17, Festival of Autumn, at Nikko, great procession. 22, "Jidai Matsuri", at the Heian Shrine in Kyoto, great procession.
November: 15, "Shichigosan", or day when children make visits to shrines.
December: 17, "On Matsuri", at the Kasuga Shrine, at Nara; procession of people dressed in ancient armour and costumes. 17-18 and 29-30, "Toshi-no-ichi" or end of year markets, in Tokyo. 17-18, at the Asakusa Temple; 20-30, at the Torigoe Shrine.

and on public holidays. It takes an airmail letter 3-4 days to reach Western Europe.

Airmail letters to Europe cost 100 yen, within Japan 20 yen. Telegrams can be sent from all post offices and hotels, they cost 192 yen per word.

Newspapers: There are four English-language daily papers: "The Japan Times", "Asahi Evening News", "Mainichi Daily News", "The Daily Yomiuri".

Police Stations: They are usually to be found near railway stations; there are police posts near busy road junctions. Japanese policemen are universally polite and helpful to visitors, they will often find you an interpreter.

Chemists: Japanese pharmaceutical products are of excellent quality and some foreign brands are made in Japan under licence. As in the West, many medicines are available only on a doctor's prescription.

Medical services: If necessary, your hotel will put you in touch with an English-speaking doctor or dentist.

Emergencies: To call the police dial 110; for an ambulance or the fire brigade dial 119.

Good advice and good manners

The rules of good behaviour are often quite the opposite in Japan to what seems correct to a European. A newly-arrived visitor runs the constant risk of appearing rude and, although the Japanese are extremely generous and tolerant in their attitudes to foreigners, there are a few gaffes which it is better to avoid. The Japanese are often rather withdrawn, they envy Westerners their extrovert sociability. But the visitor will do well to preserve a certain restraint, too much show of emotion and too much talk are not well thought of by the Japanese.

If you are introduced to a Japan-

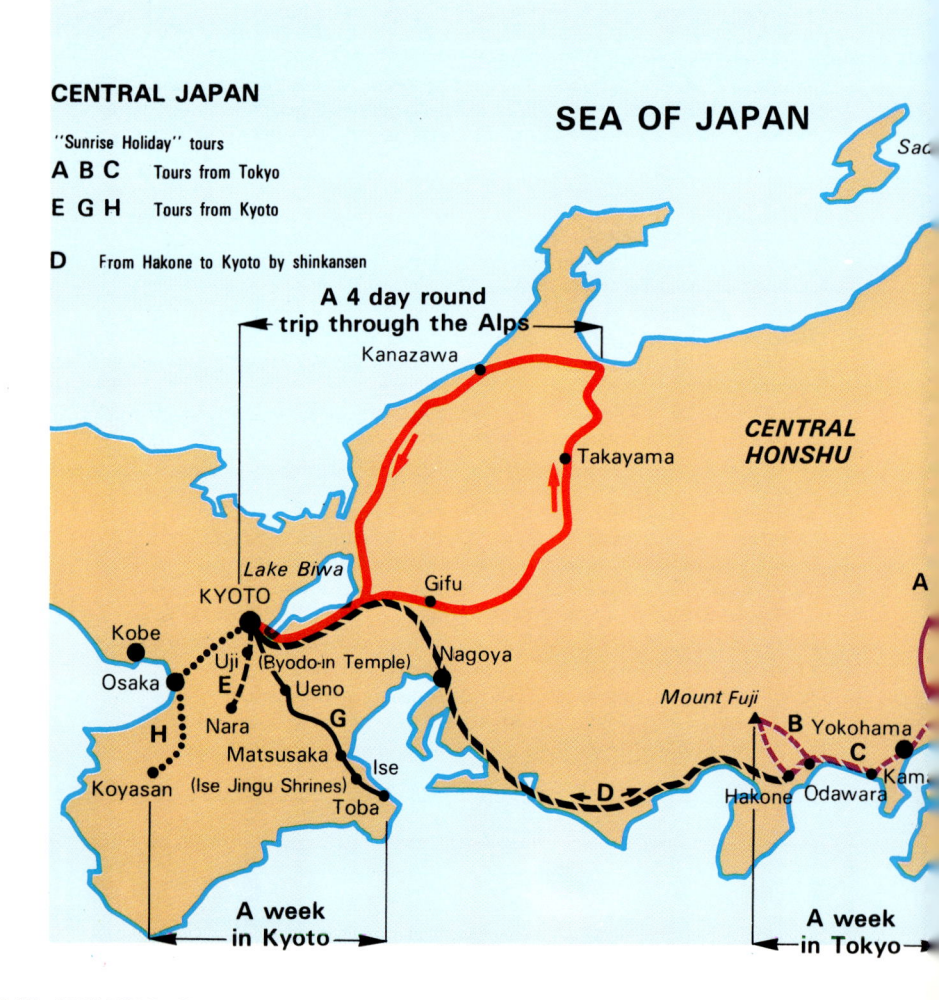

CENTRAL JAPAN

"Sunrise Holiday" tours
A B C   Tours from Tokyo
E G H   Tours from Kyoto
D   From Hakone to Kyoto by shinkansen

ese couple you should shake hands with the man, then turn to the woman, bow slightly, and wait for her to offer you her hand. Before paying a visit to a Japanese home you should call at a florist's and ask advice about the sort of flowers you should take. If your friends' flat is furnished in traditional Japanese style you will be expected to remove your shoes on entering. At table you must do your best to sit on your heels and say complimentary things about the food. Your host will reply that his "clumsy wife" has not prepared dishes worthy of the "honourable foreign guest"—but the said wife will be delighted all the same. When you are not actually using your chopsticks you should place them on the charming porcelain chopstick rest in front of you. If you are offered a second helping after having already partaken generously of sukiyaki and rice, you must politely refuse; to accept would be to indicate that the meal was inadequate.

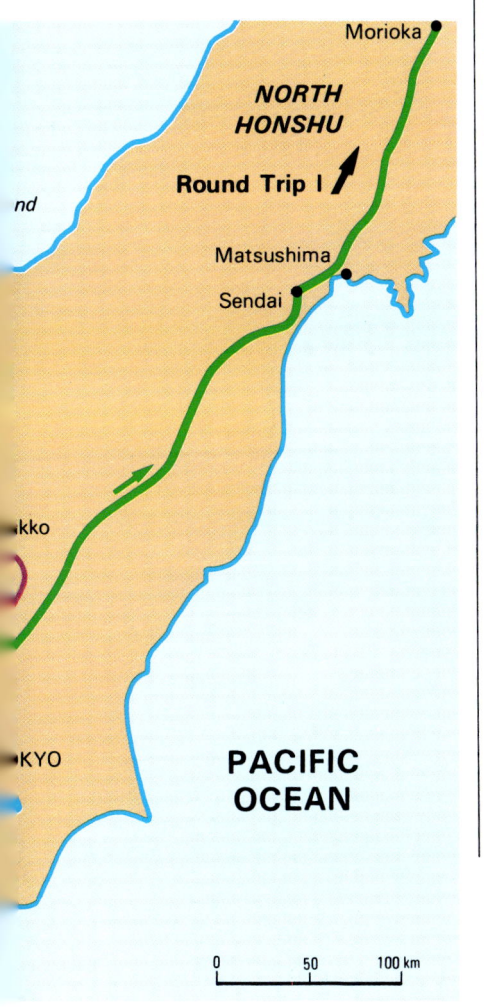

Conversation should be conducted in a low voice, with few personal references, mentioning perhaps, your journey and the lovely places you have visited in Japan. It will be particularly appreciated if you ask courteously about your hosts' children and their studies. The Japanese are immensely proud of their children and will be delighted to tell you all about their progress.

Little presents are much appreciated (see: *The art of present giving*). Your Japanese friends will be delighted to receive little presents from you—souvenirs of your travels in Japan, a box of dried seaweed perhaps, or a some local biscuits or sweetmeats.

If you are invited to spend a night in a Japanese home then you will happily sleep on "futon" (mattresses laid out on the floor) as if you have done so all your life. You will readily accept an invitation to take a bath, and will be sure to make it a good long one—a sign of your appreciation. The Japanese love being photographed; if you happen upon a wedding in your hotel the bridal party will be delighted for you to take photographs—after a polite and smiling request to be allowed to do so.

In the milling crowds on the underground, on trains or in buses, you will be stared at uncomprehendingly if you try to allow women to go first. It is thought quite normal to elbow one's way onto a crowded train. But don't be surprised to see children sitting while adults stand; in Japan the child is king.

It is considered very rude to be late for an appointment. The Japanese have a great respect for other people's work and a slight delay may well cause professional inconvenience. Finally, visitors must always remember that a smile is indispensible, however difficult their predicament may be...

accommodation

■ The Japanese are born travellers. Although they rarely take long holidays, they love going away to the mountains for a long weekend—to ski in the winter and enjoy some cool air during the summer. The seaside is very popular too, as are the great national beauty spots, artistic centres, temples and attractive old towns and cities. So there is no shortage of hotels in Japan, even in the remotest parts of the country.

Some Japanese remain faithful to the "ryokan", the traditional inns; others have acquired a passion for the latest "Western style" hotels; many establishments combine elements of both. All three types exist in three categories—luxury, medium and small.

If you travel with an organised tour you will be comfortably lodged, usually in a Western style hotel. Sometimes you may feel that your hotel lacks "local colour" and wonder why you have come so far to find something that you can find quite easily in Britain, Australia or the States.

If you are travelling independently perhaps it is a good idea to choose a Western style hotel; your journey itself is quite an exhausting business and you will have spent your day encountering plenty of novelties of one kind and another. Travel agents (JTB is highly recommended) will make hotel bookings for you. It is not advisable to arrive in a place without having booked in advance. In Tokyo, Kyoto, and Osaka there seem to be congresses going on throughout the year and at least three quarters of the hotel accommodation is permanently booked. The main tourist centres are usually crowded too, even modest hotels are likely to be full up.

The luxury or grand hotel, places like the "Imperial" in Tokyo or the "New Miyako" in Kyoto, are vast multi-storey buildings with hundreds of rooms. Scrupulously clean, somewhat impersonal, with polite and elegantly turned-out staff, their rooms are Western style, with all possible gadgets including eight-channel television. They also have Japanese restaurants serving, "tempura", "sukiyaki" and "sushi", as well as European and Chinese ones; enormous rooms for weddings and banquets; hairdressers, a shopping arcade in which you can often find French and Italian couture clothes, as well as a high class jewellers. Such hotels are not at all to be despised; the Japanese take great pleasure in patronising them, especially on special occasions, whilst their shopping arcades are a favourite among Japanese women.

It is well worth strolling round one of these luxury hotels during the spring or autumn, on Saturday or Sunday, at about midday; this is the time for weddings. There will be splendid parades of magnificent kimonos when the bridal procession arrives and the curious visitor may be able to get a glimpse of the banqueting hall with its piles of ritual wedding presents. Later, in the shopping arcades he will see some of these elegant kimono-clad figures taking a keen interest in a display of dresses by Courreges.

Several categories of hotels

The medium category hotels have all the basic facilities of the luxury ones but they are not so stylish. Instead of a shopping arcade there will probably be just one shop, selling everything from razors to radios and souvenirs. Often their restaurants will offer a combination of Japanese and Western cooking. They correspond to good medium range hotels in the West and their prices are much more reasonable than the grand hotels.

Finally there are the small western style hotels; frankly, it is better to choose a "ryokan"—especially if you have a "good address" from a Japanese friend or are keen to try something entirely different.

"Ryokan" are traditional Japanese inns. Their prices may seem high, but they include the room, the evening meal and breakfast. They are built of wood and are expensive to construct and to run, so there are fewer and fewer of them. They are usually most beautifully situated.

The "Ryokan"

Guests normally remove their shoes on entering a ryokan; slippers are available on loan from the kimonoed staff. A smiling hostess will take you to your room. This consists of a little entrance—where you leave your slippers—with the bathroom opening off it, and, up a few steps, a classic Japanese bedroom, with tatami on the floor and cupboards concealed behind fusuma; it is all very light and pleasant, with an "ikebana" in the "tokonoma", a "kakemono" (see: *Glossary*), a low table, two cushions and a little dressing table with an oval mirror. Sometimes there is a little verandah with a couple of armchairs—an invitation to relax and look at the view or the lovely garden.

As soon as you enter your room hot green tea will be brought in. Then, if it is the end of the afternoon, you will take your bath. You will find that the bath itself is sunken into the floor; you actually get into the bath only after you have carefully soaped yourself all over and then rinsed with water from the little wooden bucket provided. Your hostess will have been careful to fill you bath with only "lukewarm" (40 °C.) water, as you are a foreigner and not used to the near-boiling temperatures in which the Japanese take such delight. Then, dressed in your "yukata" (cotton kimono, provided by the inn) you will take your dinner, normally in your room. The meal will consist (unless you speak Japanese or have come to one of the rare ryokan where English is well understood and have thus been able to order a particular meal) of a number of different dishes, charmingly arranged on a tray.

After dinner your table will be pushed up against the wall and the "futon" (beds, or rather mattresses) will be brought out from their cupboards and laid out on the tatami, and your night light will be lit for you. You can then either go for a stroll, in your yukata, round the inn's shopping arcade, say, or settle down to watch a little television before you go to sleep. The bed consists of a soft mattress, an eiderdown in a cover, and a small pillow. It may seem strange at first to be sleeping virtually on the floor, but these beds are remarkably comfortable and you are likely to have an excellent night's rest.

It is probably a good idea to spend one's first ryokan nights in a good quality establishment. Simpler ones exist, and they are of course much cheaper, but it is better to get the hang of things in a really well-equipped inn (possibly English-speaking too) first. Many visitors find staying in ryokan one of the most enjoyable features of a holiday in Japan.

There are hotels which combine both Western luxury and the traditional features of the ryokan—the Suginoi Hotel at Beppu, for example (see special entry). In such hotels you can take your pick of both worlds—sleep in a Western-style bed or on futon, dine in your room or in a restaurant, choose either a Western or traditional Japanese breakfast. Such hotels often have one or several swimming pools as well as separate public baths for men and women. □

SHIKOTSU NATIONAL PARK

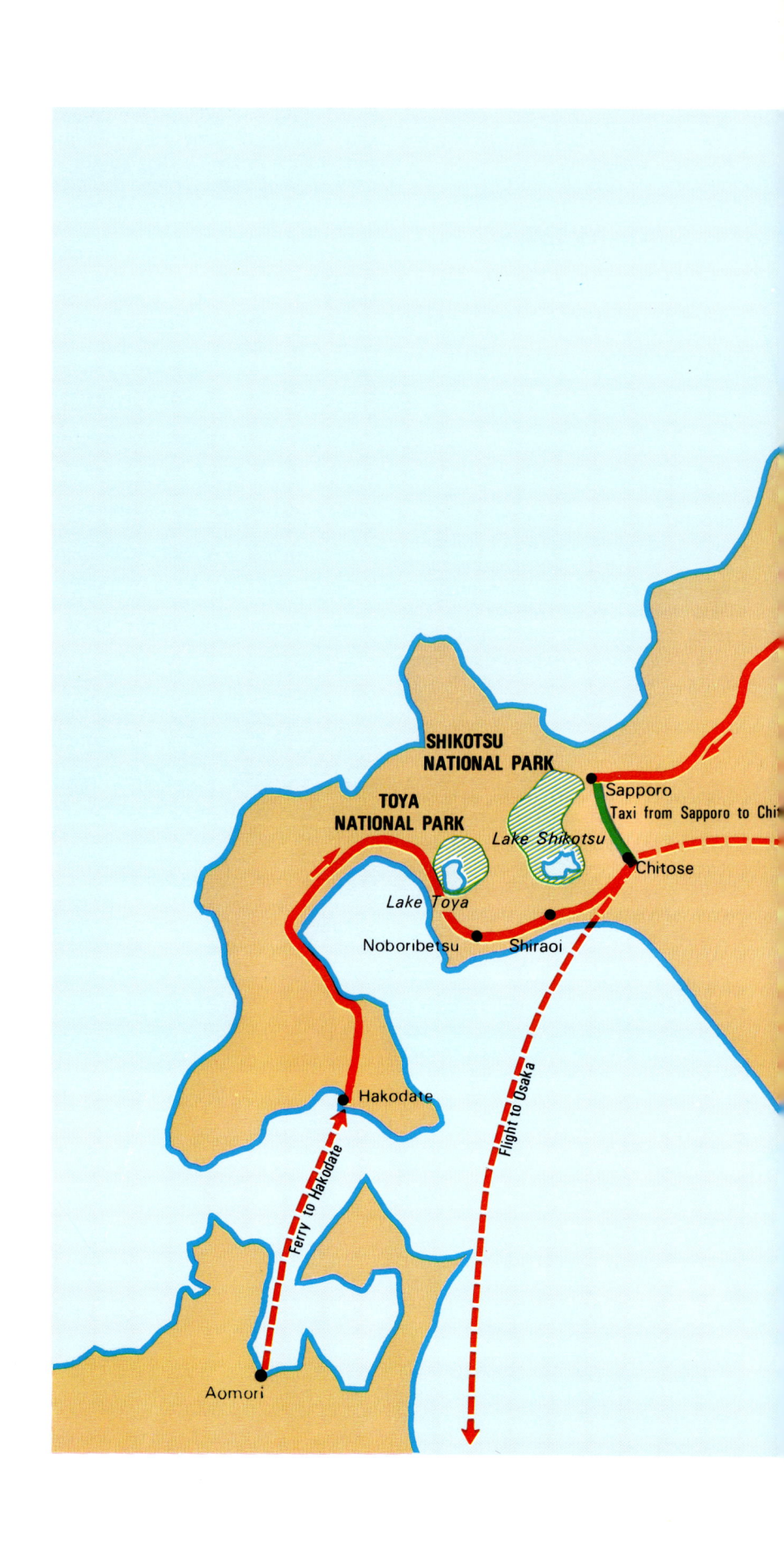

TOYA NATIONAL PARK

Sapporo
Taxi from Sapporo to Chi[tose]
Lake Shikotsu
Chitose
Lake Toya
Noboribetsu Shiraoi

Hakodate

Ferry to Hakodate

Flight to Osaka

Aomori

HOKKAIDO: 9 day tour

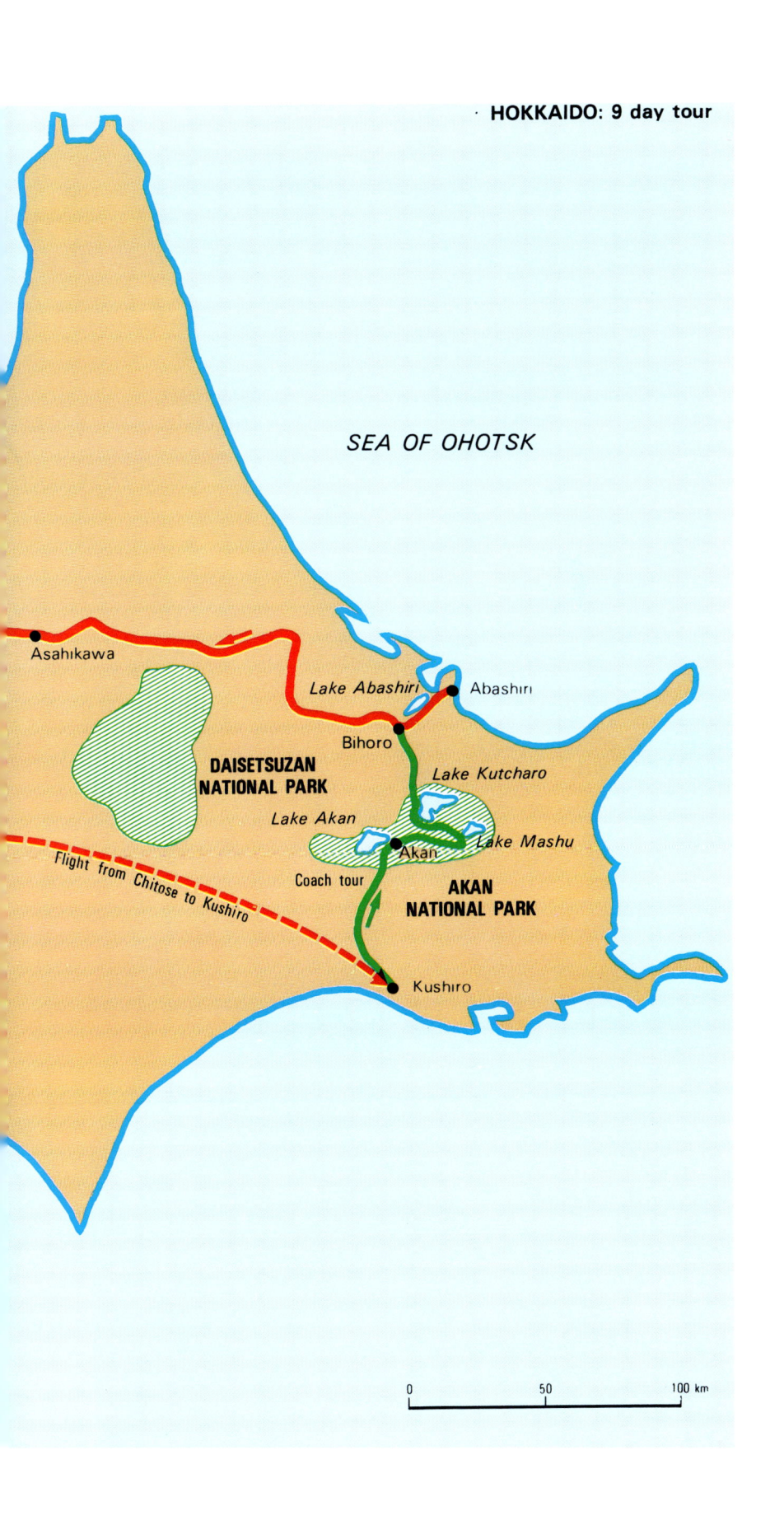

SEA OF OHOTSK

Asahikawa

Lake Abashiri — Abashiri

Bihoro

DAISETSUZAN NATIONAL PARK

Lake Kutcharo

Lake Akan

Lake Mashu

Akan

Coach tour

AKAN NATIONAL PARK

Flight from Chitose to Kushiro

Kushiro

0   50   100 km

JAPANESE JOURNEY 223

japanese food

■ In our chapter on *The Japanese Table* we attempted to convey something of the Japanese attitude to eating. We now propose to discuss some of the Japanese dishes that visitors will wish to sample.

Europeans arriving in Japan should try to forget their own native dishes—if only because European cooking in Japan tends to be very expensive and is not always very good. But there are better reasons; Japanese cooking is both varied and interesting. Just go out into the streets and look! Although the Japanese are economical they frequently take their meals in restaurants—so restaurants abound. They often display in their windows wax models of the dishes they serve—often remarkable works of art in themselves in which every vegetable, every piece of meat is enticingly reproduced. When you enter the restaurant, ordering is a simple matter—you simply point to the dish you fancy; it will appear in due course, an exact counterpart of its model. Japanese restaurants tend to be restful and, except during the rush hour between midday and one o'clock, uncrowded. As soon as you sit down a waitress will bring you green tea and hot towels on which to wipe your hands.

Fish, eaten both raw and cooked, is a staple of the Japanese diet, and vegetables are another; meat is expensive and appears only rarely. The Japanese do not eat mutton or goat and little cheese is available. The great dishes of Japanese cuisine are: "sukiyaki", "sushi", "sashimi" and "tempura". They are usually served in specialised restaurants; a place serving "sukiyaki" will not serve "tempura", and vice-versa, so couples with diverging tastes are sometimes at pains to decide where to eat.

"Sukiyaki" (pronounced "skiyaki") is usually a favourite with visitors to Japan. Its wax model consists of a handleless pan in which various vegetables are piled—leeks, onions, mushrooms—together with thin slices of beef streaked with fat and transparent noodles. The diners seat themselves on cushions around a low table, onto which the waitress places a small heater. Then, using sake and soja, she cooks the "sukiyaki" in front of them, serving them with the various items as they cook. It is customary to dip each mouthful into a bowl of beaten egg before putting it into one's mouth. With "sukiyaki" it is normal to drink green tea or hot sake. "Sukiyaki" for two in a modest Tokyo restaurant will cost about 4,000 yen.

To please the taste

"Tempura" is a dish of Portuguese origin that came to Japan during the 16th century. The Portuguese did not eat meat during the hot summer months so they replaced it by fried fish. They greatly enjoyed the large Japanese shrimps. The Japanese took over the Portuguese recipe and improved upon it. "Tempura" is a highly refined dish. The oil in which it is fried is a carefully judged mixture of peanut and sesame oil; every cook has his own recipe for it. The chef prepares the "tempura" in front of his customers and serves it on the bar, close to the kitchen. A "tempura" meal has from six to twelve items: large shrimps, pieces of fish "kisu", squid, aubergines, mushrooms, beans, scallops, peppers, mussels, sweet potatoes... and chrysanthemum leaves! Each item is dipped in a batter made of egg, flour and iced water, then plunged into the oil at a temperature of 190 °C. The batter swells up and the whole thing is very light and highly digestible. Restaurants specialising in "tempura" display an exact model of it in their windows, so prospective customers can see whether they serve mushrooms or peppers. Western visitors may find they need two servings of "tempura" to stay their hunger; even then it's not an expensive dish, at about 800-1,000 yen a time.

The visitor may be puzzled by the sight, in shop windows of wooden boxes of what look like little cream cheeses surrounded by dried seaweed; these are known as "nori mari"; one or two are often eaten after a meal of "sushi", they are also a favourite pic-nic food. The shops that display them sell "sushi" (rice with vinegar) and "sashimi" (slices of raw fish). These are perhaps not dishes for the newcomer to Japanese food—especially if he or she is squeamish about eating raw fish. But they are tastes well worth acquiring; these dishes are nourishing and delicately flavoured. "Sushi" consists of a bed of rice with vinegar, on this are placed slices of omelette, salad, horseradish or mushrooms, and raw fish, tuna, or squid, or prawns. The filling is then wrapped round in vinegar rice and the whole thing is then cut up and served in slices, presented in attractive wooden boxes. Other kinds of "sushi" are stuffed with red caviar or shrimps, or surrounded with fillets of tuna or bass. There is a great variety of sushi. In some restaurants they are prepared by a chef, and served straight to customers who sit up at a horse-shoe-shaped bar to eat them, often dipping them into soy sauce first. A small plate of sushi costs about 300 to 400 yen, the price varies according to the quality of the fish used.

"Sashimi" are all sorts of raw fish fillets, eaten with a piquant sauce of soy and horseradish. Such fish must be very fresh, non-frozen, and must never be touched with the fingers during its preparation for the warmth would spoil its flavour. The best sashimi are made with thick slices of tuna. White-fleshed fish are served in very thin slices and often elaborately arranged. Thus, "fugu"—a fish that has to be cleaned extremely carefully as it contains a sac of virulent poison—which the Japanese delight to eat during the winter months, is often set out so that its delicate transparent flesh looks like a crane in flight. Thus beauty is combined with danger (for there is always a slight risk), a combination which appeals very much to the Japanese; "fugu" may not have much flavour but it is a highly esteemed dish.

SAKE: A VERY DRINKABLE WINE

Sake is rice wine. Its origins go back to the earliest years of Japanese history, to the Yayoi Period. It is a fermented drink, made in a rather similar way to beer. When served cold, sake is known as "reishu"; it is sold everywhere, even on railway platforms and in trains. It is even better drunk hot—heat brings out its flavour—from little porcelain cups known as "choko". Sake is the ideal drink to take with "sukiyaki", "sushi", and "sashimi". It is available in different strengths; the Japanese vary them according to the dish being served. The annual consumption of sake in Japan is 13 litres per head.

As well as these special restaurants for sukiyaki, tempura, sushi and sashimi there are many others, small places perhaps, where the Japanese go for a simple lunch, or supper around 6 o'clock. They offer simple typically Japanese dishes such as "tougatsu", "shabu shabu" or "mizu-taki"—all popular with Westerners —or Indian or Italian food presented in a Japanese way. The wax models in the window show a choice of soups (often extremely thin) and "soba", a popular dish, eaten by the Japanese at all hours of the day. It consists of a large bowl of soup with noodles, a few leaves of salad and two fried prawns floating in it. It's easy enough to drink the soup, but it's more difficult to trap the slippery noodles on your chopsticks! Soba is frankly not the most exciting dish to a Western palate. These little restaurants also serve rice omelettes with tomato sauce, curry and rice, and often spaghetti served with a tomato sauce that seems to combine both East and West. Such dishes are not expensive (500-600 yen) but are frequently not very filling; you can always order two.

In the larger cities there are often attractive "Korean" restaurants. In some of them you dine (without your shoes) sitting on cushions on the tatami, at low tables on which there is a little grill. The waitress brings a number of plates of raw meat—chicken, pork, liver, gizzards—which you grill yourself and eat with peppers and salads. Such restaurants tend to stay open late. Their prices are around 1,500 to 2,000 yen per head for a meal.

In Honshu the big cities tend to have almost as many Chinese as Japanese restaurants. To anyone knowing Chinese cooking they will seem perfectly familiar. They also display wax models—of Cantonese rice, sweet and sour pork and other classic dishes. They offer good and varied menus of Peking, Cantonese and Shanghai food; their prices tend

THE INLAND SEA AND KYUSHU

Eight days in Kyushu

to be around 1,000 yen.

All larger hotels have restaurants serving sukiyaki, tempura, sushi and sashimi. Here the dishes are always well cooked and attractively served; helpings are generous but prices are much higher. A good sukiyaki for two may cost 5,000 yen, while a main course and a sweet in the same hotel's Chinese restaurant will cost more than 1,500 yen per person.

In the traditional inns or ryokan the evening meal and breakfast is always included in the price of the room. In the larger ones it is possible to order a meal in English, but in small country ryokan the visitor will have to take the traditional Japanese meal without question. This will be very much the same as the Japanese eat in their own homes (the average consumption is around 2,100 calories per day) and will almost certainly contain fish and one or two new items. The evening meal will be brought to your room at 7 o'clock and is likely to consist of a bowl of clear soup with shellfish, a long dish with three good slices of raw tuna (sashimi), decorated with slices of carrot, an attractive cream caramel (which turns out to be salted and tastes slightly of fish) and a lacquer dish of rice and a small helping of cooked fish. Breakfast is likely to be rather similar unless the ryokan goes in for some Western dishes.

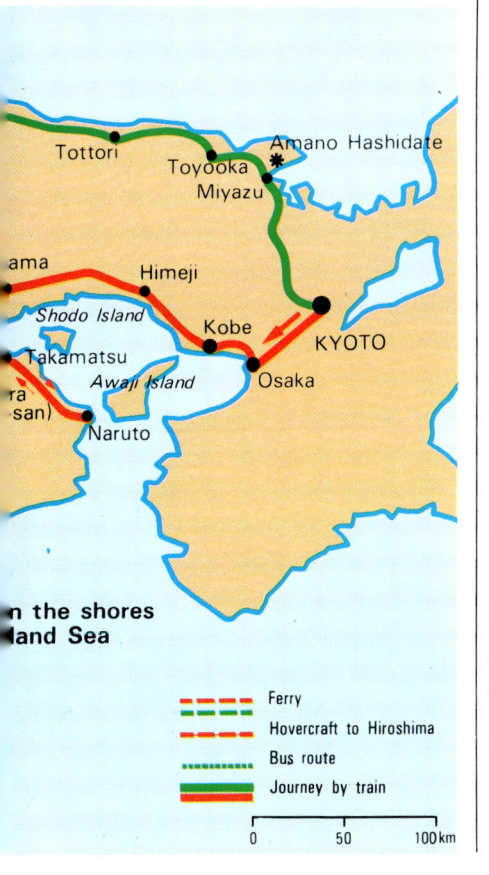

Sometimes, during long journeys, meals will have to be taken on the train. These are very well organised. On the "Hikari" shinkansen there is always a dining car which serves a small selection of well-prepared Japanese and Western dishes. In almost all stations it is possible to buy little packed meals, hard-boiled eggs, fruit, and bottles of sake, beer and fruit juice as well as green tea. Packed meals can also be bought on the trains themselves; they sometimes contain ten sushi (rolls of vinegar rice with seaweed) or rice and fish or meat, or a slice of omelette. These meals are served warm and eaten with chopsticks; they are good, and attractively presented.

Obviously there is more choice of food in the larger cities than in the remoter country districts or small spas, but there are often pleasant eating places and interesting regional dishes to be tried (see, for example, the special entry: *A little restaurant in Hakodate*).

All over Japan the visitor will be most charmingly served by waiters and waitresses, all most willing to help in the choice of dishes, and explain how to use chopsticks or the correct way to drink sake. In a little village your fellow guests at an inn will often be delighted to show you how an unfamiliar dish should be eaten.

If you are really homesick there are always the Western restaurants in the big hotels, where they served French, Italian and other dishes, often slightly tinged with Japanese elements. Their prices tend to be high; a simple meal can easily cost 5,000 yen apiece, though hotel coffee shops are often much cheaper.

What to drink

Green tea is served and drunk at all hours (see special entry), there is excellent Sapporo beer, wine is very expensive if it is imported and the local wines are not very good. Sake is served cold (when it tastes slightly of eau de cologne) and hot (very good indeed).

Chopsticks

Restaurant chopsticks are made of wood, known as "waribashi", and presented wrapped in paper. They are used for one meal only. When you unwrap them you will find that they are still joined at one end; they split very easily. The Japanese rub their "waribashi" together before using them, to make them smooth and to get rid of any splinters. After the meal they should be replaced in their paper wrapper. It is not considered good manners to cross your chopsticks on your plate or to leave them sticking into your food while you talk. If you are invited to a Japanese home, special chopsticks—of lacquer, bamboo or ivory—will be used in your honour; they should be carefully laid down near your bowl when you have finished. Before taking "pikkels" (salted vegetables, served with a bowl of white rice, "tsukemono") you should dip the tips of your chopsticks into your green tea. □

THREE KINDS OF TEA

■ "*Usu cha*" is a very delicately-flavoured tea. It consists of finely ground young shoots, which dissolve when boiling water is poured on them. When beaten up with a tea whisk a pale green infusion results. "*Usu cha*" is the variety used in the traditional tea ceremony.
The second kind of Japanese tea is "*o cha*", a green tea made by dipping the tea leaves in boiling water and then drying them in a warm place. It is drunk anywhere and everywhere, all the time, just like water.
The third kind is called "*kocha*". The leaves are allowed to ferment, which gives the tea a red colour and a quite distinctive flavour.

sport

Sumo

Sumo wrestling tournaments are held six times a year and each one lasts for a fortnight. They take place in Tokyo in January, May and September, in Osaka in March, in Nagoya in July and in Fukuoka in November. Radio and television have made the sport popular throughout the country; even a modest visitor, quietly eating his crabs in a little bistro in Abashiri in the Far North, can follow the bouts on the small screen, with helpful comments from his fellow diners.

The origins of this traditional form of combat are lost in the mists of history, according to legend it is about 2,000 years old. It seems to have grown out of rituals that took place in the courtyards of Shinto shrines. By the 8th century annual bouts were taking place early in the year, in the presence of the Emperor. The present rules of Sumo were codified during the 16th century.

The wrestlers are professionals. They are huge men, weighing as much as 130 kilos. Late in the afternoon, before a capacity audience the two sons of the wrestlers walk into the ring, ahead of the famous super-champions. They are dressed in a kind of apron and their hair is done up into a little chignon. They bang on the ground, to drive away any evil spirits, then withdraw. The first pair of wrestlers then enter the ring, throw down a few handfuls of salt in purification, start limbering up and eye each other carefully, trying to spot possible weak points in their opponent. They are naked except for a jock-strap—often quite elaborate —knotted below the waist.

When the ritual four minutes' wait is up, these two great mounds of flesh hurl themselves at each other, each attempting to use one of the fifty authorised holds to throw his rival onto his back or make him move outside the strictly delimited wrestling area. A bout rarely lasts more than two minutes. The referee is dressed like a Shinto priest. The judges, all former wrestlers themselves, are seated round the ring. Their decisions are accepted without question. During the rest of the year the "sumo" champions go on tours of the provinces.

Martial Arts

When the first Shogun, Minamoto Yoritomo, set up his government at Kamakura, during the 12th century, he surrounded himself with valiant warriors. They disdained Kyoto and the frivolities of Court life there and followed a much more rigorous lifestyle, codified later as "Bushido" or "The Warriors' Path". From a very early age, the future warriors learned "kuyodo" (archery), "jiu-jitsu" and "ken-jutsu" (fencing).

During the 13th century the samurai enthusiastically took up "Zen", with its twin doctrines of the liberation of the spirit through meditation and severe bodily discipline. The practice of the Martial Arts was a way to achieve enlightenment or "Satori". At the beginning of the Meiji period (see: *Japan through the centuries*) it was forbidden to carry arms—they were seen as a symbol of feudalism—and the Martial Arts declined. They revived after the turn of the century, however, and continue to flourish today. The most popular branches have become competitive sports, like judo, while others are still taught in traditional schools, according to the ancient rules.

The practice of any Martial Art demands great respect by the student for his teacher, an attitude of non-violence, a calm containment of energy, and control of breathing. The novice has to learn how to "breathe with the stomach", held in the Far East to be the vital centre of energy.

"Kudyo" (archery), formerly reserved for an elite, has thousands of practitioners today. Its primary object is to be a means of enlightenment, of seeing the world as a unity, of abolishing the duality between the "self" and the "non-self". It can be considered as an almost spiritual exercise. Apprentice-ship in archery lasts several years. The Japanese bow is more than two metres long and the arrows are long and heavy. The archer shoots at fixed targets, either standing or on horseback. The peak of achievement as a bowman, they say, is to be able to hit the target blindfold! In Kamakura there is a colourful historical festival held every year (14-17 September) at the Tsurugaoka Shrine. A line of targets is set up, with cups filled with confetti. The archers, dressed in 12th century costume, compete one by one. The man who shatters the most cups is the victor of the day.

The art of "kendo" or "The Way of the Sword", which dates from the 13th century, consists of unsheathing one's weapon so quickly that one's opponent is taken off guard. Despised after the war, kendo is gradually regaining popularity. The sword has been replaced by four bamboo rods, joined end to end, lashed together with leather thongs. A helmet must be worn as head-blows are allowed. Some clubs still teach a form of fencing with two swords, "nitto", and "nagitaka" in which a wooden weapon, two to three metres long, is used.

During the 15th and 17th centuries the islands of Okinawa were occupied successively by the Chinese and the Japanese. The local inhabitants were forbidden to carry weapons, so to defend themselves they adapted a form of hand to hand fighting they had previously learned from the Chinese. This form of fighting, known as "karate" is extensively taught today. It is practised in special halls, on tatami (mats); the participants wear a pyjama-like garment, with a belt whose colour varies according to their strength. Karate demands great coolness and self-control. The blows should be swift and decisive. The classic attack is a blow from a clenched fist or with the back of the hand, using one, two, or three fingers—usually aimed at the abdominal muscles of one's opponent. Kicking, with the ball of the foot and the heel, is also allowed.

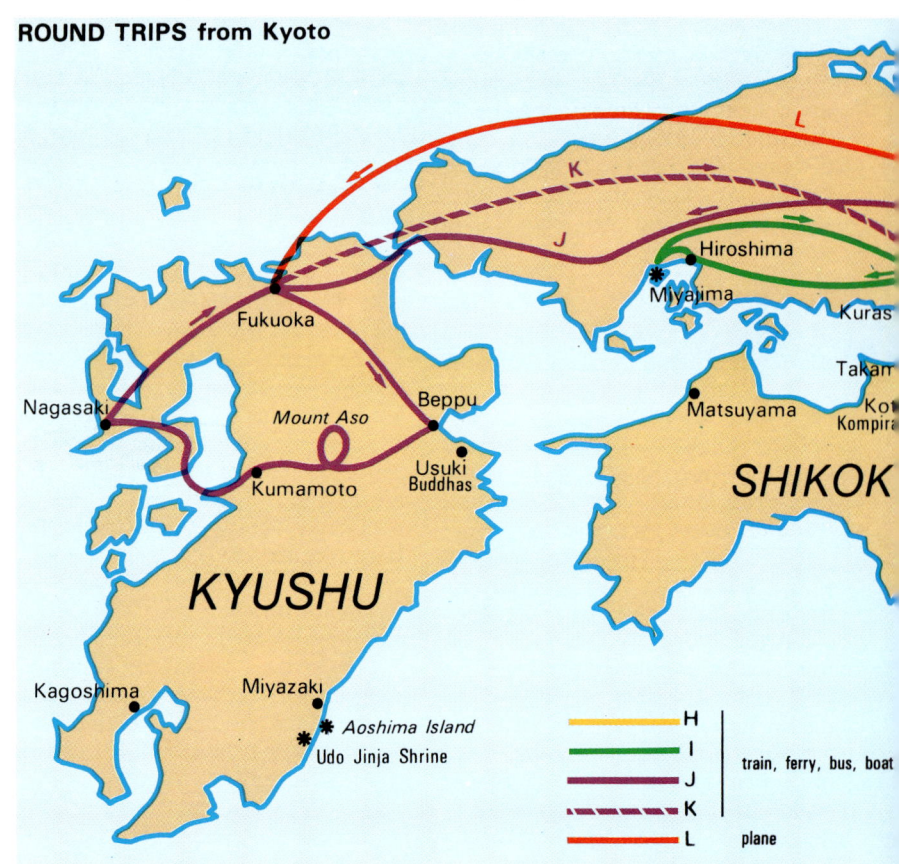

ROUND TRIPS from Kyoto

The art of "aikido", "the way to harmony in unity", goes back to the 12th century. "Mind and body must achieve such harmony that movement becomes as natural as breathing. All gestures thus partake of the nature of breathing or of music" (Michel Randon).

"Judo" which developed out of the ancient art of "jiu-jitsu" has become a very popular competitive sport. It was included in the Olympic Games for the first time in 1964, and again in Munich in 1972. The first international judo championships took place in Tokyo in 1956. Every year, the National Judo Tournament creates enormous interest throughout Japan.

Other sports

Apart from these traditional sports the Japanese are great devotees of baseball, which is played in schools from an early age. Ski-ing is becoming more and more popular too. It is estimated that there about eight million skiers and skaters in Japan today. The mountains attract more climbers every summer; the Japanese have acquired a world-wide reputation for their exploits in this field.

Golf

The visitor to Japan may be surprised to see an enormous range of golfing accessories on sale in the shops. Only twenty years ago this sport was considered the preserve of the upper classes and the very rich, plus a few businessmen and politicians. Today it is played by all classes, in their millions. This phenomenon is an aspect of the prosperity Japan achieved during the 1960's. People were wealthier and had the means to escape from the crowded, polluted cities. There was an increase in sports' reporting on television and golf championships were immensely popular with viewers who saw the sport as an ideal form of relaxation —and one that they could now themselves afford. The popularity of the sport grew rapidly.

Most courses were some distance from the city centres, however, and businessmen couldn't easily go out to practise after work. So practise halls were built in the cities; almost every district now has one. Their characteristic half-circular shapes are easily recognised; hundreds of aspiring golfers practise there every day. And not only there—it is quite common to see the man next to you in a bus queue lifting his arms to practice his golf-swing, nobody pays the slightest attention. Early on Sunday mornings you can see hundreds of eager golfers, laden with clubs and golf-bags, setting off by train for one of the 700 courses that Japan now possesses. □

a thousand souvenirs

■ It costs 1,000 yen to air-freight a 1 kilogramme parcel and it takes about a week for it to reach Europe—up to 20 kilogrammes can be sent at a time. This is a fact well worth bearing in mind for the 20 kilo luggage limit imposed by the airways becomes increasingly irksome. The little shops in Japanese hotels are quite used to freighting parcels all over the world.

A week before returning home then, you can put all your Japanese purchases into a suitcase, take it to the airport (by the hotel bus), and send it off from the freight office. It's slightly expensive, but it would be a pity not to take away plenty of souvenirs.

Japanese radios, cameras and cine-cameras have a worldwide reputation these days. There are so many new brands on the market that it's a good idea to look in all the specialised shops—in the hotel arcades as well as in the shopping streets—for foreigners can buy such things tax free, though it is advisable to find out about the customs duties payable at home.

Cultured pearls are a temptation in Japan. There is such a selection at the jewellers' that it's often very difficult to decide between items —necklaces, bracelets and rings—and colours—"orient", blue and black —and prices. It is essential to do the rounds before making a final choice.

The grandest kimonos are fabulously expensive. But in every town there are kimono shops where they will make them to measure in attractive silks from Kyoto and Kanazawa (small patterns) or Okinawa (larger, slightly exotic patterns). A kimono takes about 12 metres of cloth (the silks are normally 37 cm wide) and prices range from 50,000 yen (roughly 400 yen to the pound). In the shopping arcades of the larger hotels you can buy pretty kimono-dressing gowns, at reasonable prices.

The "yukata", a sort of cotton kimono that the Japanese wear at home in the evenings, is an attractive garment. Yukatas are provided for the use of guests in ryokan; you can often buy similar ones at the hotel shop (3,000 to 5,000 yen). The yukata is pleasant to wear and easily laundered—you will probably find yourself catching the habit of wearing one at home like the Japanese.

In Kyushu, at Nagasaki particularly, there are craftsmen who make all sorts of tortoise-shell objects—everything from model ships to combs. They also use it with red coral to make jewellery.

It would be a pleasure to furnish one's house in Japan. The Japanese vary their table settings according to the seasons; there is a vast choice of china and porcelain at reasonable prices. In the big stores you can find delightful tea and rice bowls, and sake glasses, packed in attractive wooden boxes. Chopstick rests are another possible souvenir; they come in many forms—crescent moons, fans, fruits and vegetables and fish —and have the advantage of being light and easily packed.

Lacquerwork is lovely but expensive.

Each region has its speciality: from wooden dolls to all sorts of objects made of bamboo. Friends at home will be delighted with such typically Japanese items as boxes of sea-weed, little dried fish, perhaps a charming octopus (a speciality of Amano Hashidate) or a bottle of sake.

A thousand souvenirs await the visitor to Japan. It only remains to wish him "itte irasshai"—"an honorable journey and an honourable return!" □

a few words of japanese

NUMBERS

1	ichi, hitotsu	20	ni ju
2	ni, futatsu	21	ni ju ichi
3	san, mittsu	30	san ju
4	shi, yon, yottsu	80	hachi ju
5	go, itsutsu	90	kyu ju
6	roku, muttsu	99	kyu ju ku (or kyu)
7	shichi, nana, nanatsu	100	hyaku
8	hachi, yattsu	200	ni hyaku
9	ku, kyu, kokonotsu	1000	sen, issen
10	ju, to	2000	ni sen
11	ju ichi	1000000	kyaku man

TIME

year	nen, toshi	month	tsuki
last year	sakunen	January	ichi gatsu
next year	rainen	February	ni gatsu
this year	kotoshi	March	san gatsu
season	kisetsu, shiki	April	shi gatsu
spring	haru	May	go gatsu
summer	natsu	June	roku gatsu
autumn	aki	July	shichi hatsu
winter	fuyu	August	hachi gatsu
week	shu	September	ku gatsu
day	hiru	October	ju gatsu
a day	ichi nichi	November	juichi gatsu
morning	asa	December	juni gatsu
this morning	kesa	this evening	konban
noon	o hiru	afternoon	gogo
Monday	getsu yobi	night	yoru
Tuesday	ka yobi	tonight	konya
Wednesday	sui yobi	today	kyo
Thursday	moku yobi	yesterday	kino
Friday	kin yobi	the day before yesterday	ototoi, issaku jitsu
Saturday	do yobi	tomorrow	ashita
Sunday	nichi yobi	hour	jikan
a day	ichi nichi	minute	fun

THE JOURNEY

customs	zeikan	frontier	kokkyo
consulate	ryoji kan	passport	ryoken
embassy	taishi kan	vaccination	shuto
foreigner	gaikoku (jin)	alcohol	arukoru
man	jin	tobacco	tabako
export	yushutsu	restaurant	resutoran (Western)
the hotel	hoteru (Western)	restaurant	ryoriya (Japanese)
the inn	ryokan (Japanese)	chopsticks	hashi
bill	kanjo	beer	biru
luggage	nimotsu	steak	bifuteki
room	heya	bowl, cup	chawan, koppu
full	man in desu	knife	naifu
blanket	kake buton	spoon	supun
key	kagi	fork	hôku
maid (to call)	ojosan	water	mizu
bathroom	furoba, basurumu	hot water	o yu
telephone	denwa	chef	ryorinin (Japanese)
hello	moshi moshi	chef	kokku (Western)
mail, letters	yubin		
bed	betto (Western)	coffee	kohi

JAPANESE JOURNEY 233

IN THE RESTAURANT

eel	unagi	breakfast	choshoku asakan
chopsticks	hashi	lunch	chushoku
beer	biru	dinner	yushoku
steak	bifuteki	fruit	kudamono
beef	gyuniku	egg	tamago
coffee	kohi	omelet	omuretsu
lemon	remon	bread	pan
crab	kani	fish	sakana
shrimps	ebi	cooked rice	gohan
cake	o kashi, kêki	sandwich	sandoitchi
fruit juice	furutsu jusu	sugar	sato
lemonade	saida	tea	o cha (Japanese)
		tea	ko-cha (Indian)
tuna	maguro	meat	niku
salt	shio	wine	budo shu

TRANSPORT

airport	kuko	plane	hikoki
station	eki	bus	basu
harbour	minato, ko	boat	boto, fune
taxi	takushi	luggage	nimotsu
underground	chikatetsu	ticket	kippu
seat	seki	disembark	joriku suru
reserved seat	yoyaku seki	departure	shuppatsu
porter	akabo	embark	jôsen suru

please call a taxi — denwa de takushi wo yonde kudasai (please)
take me to the station — eki made tsurete itte kudasai
go straight ahead — massagu ni
to the right, to the left — migi e, hidari e
stop here — koko de tomatte kudasai
wait here — koko de matte kudasai
I'll be back in a moment — sugu ni kaerimasu
what time is the train for...? — yuki no kisha wa nanji ni demasu ka?
what platform is the train for...? — yuki wa nan bausen kara demasu ka?

TRAIN TRAVEL IN JAPAN

arrival	tochaku	booking office	shussatsu guchi
luggage	nimotsu	change trains	kisha wo norikaeru
ticket	kippu	left luggage	nimotsu azukarijo
class (first and		seat	seki
second)	itto, nito	reserved seat	yoyaku seki
couchette	shindai	platform n. 1	ichiban-sen
station	eki	express	kyuko
train	kisha	semi-express	junkyu
fast train	tokkyu		

where is the booking office? — shussatsu guchi wa doko desu ka?
what day are you leaving? — itsu odekake desu ka?
I am leaving on... — ni dekake masu
what time does the train leave? — yuki no kisha wa nanji ni demasu ka?
give me a first class single — yuki no itto o ichimai kudasai
give me a second class return — yuki no ofuku no nito o nimai
how much is it to...? — made ikura desu ka?
which platform is the train for...? — yuki wa nan bausen kara demasuka?
have a good journey — itte irasshai, o genki de
taxi driver — takushi, untenshu
take me to the P. hotel — Hoteru P made tsurete itte kudasai
stop here — koko de tomatte kudasai
wait here — koko de matte kudasai
please show me — wa doko desu ka
I want to go to — e ikitai desu

234 JAPANESE JOURNEY

OTHER WORDS

bank	ginko	information office	annaijo
library	tosho kan	booking office	kippu uriba
exhibition	tenjikai	garden, park	niwa, koen
police	koban	chemist's	kusuriya, yakkyoku
it's fine	ii o teki desu	it's hot	atsui desu
it's cold	samui desu	it rains, it is raining	ame ga futte imasu
it's windy	haze ga huite imasu		
sorry to bother you		ojama shite sumimasen	
excuse me, sorry		sumisamen, gomen-nasai	
yes	hai		
no	iie	where	doko
how much/many	ikura	here	koko
		there	asoko
how much is this?		kore wa ikura desu ka?	
where is it? It's there		doko desu ka? koko desu	
how are you?		gokigen ikaga desu ka?	
I am very well, thank you		arigato, genki desu	
do you speak English?		furansugo o hanasemasu ka?	
I do not understand		wakarimasen	
please		kudasai	
don't mention it		dozo	
thank you very much		arigato gozaimasu (domo arigato)	
wait a moment		chotto matte kudasai	
good bye		sayonara	
sorry to bother you		ojama shite sumimasen	
it's too much		amarini osugi masu	
it's too little		amarini sukuna sugimasu	
please show me		wa doko desu ka	
I would like to buy		o kaitai desu	
how much is this?		kore wa ikura desu ka?	
it's too dear		takasugi masu	
will you wrap it?		yoku tsutsunde kudasai	
I would like to send a telegram		dempo wo uchitai desu	
have you a room?		heya ga arimasu ka?	
by express mail, by air mail		sokutatsu, kokubin	
white	shiroi	beautiful	utsukushii
blue	ao, buru	much	takusan
grey	nezumi iro	good, well	yoi
yellow	kiiro	hot	atsui
black	kuro	cold	tsumetai, samui (weather)
red	aka	big	okii
green	midori	bad	warui
fog	kiri	open	aku
snow	yuki	small	chiisai
rain	ame	little, a little	sukoshi, shosho
sun	taiyo	nothing	nanimo
earthquake	jishin	often	tabi tabi, toki doki
typhoon	taifu	too	amarini

GLOSSARY OF JAPANESE WORDS

Ainu	aboriginal population (see: Japan through the centuries and entry for Hokkaido)	Confucianism	religion
		cryptomeria	common Japanese conifer
		daimio	lord of a province
bashi, hashi	bridge	dori	street
bento	"packed lunch" sold at stations and on trains	en, koen	garden, park
		eki	station
Buddhism	religion	furoshiki	cotton scarf, often used—knotted—to carry parcels or presents
chanoyu	tea ceremony		
cho	district (in a city)		

fusuma	sliding screen or wall		of allpowerful prime minister, exercising power in place of the emperor, an hereditary office.
futon	Japanese bed		
gawa, kawa	river		
geisha	hostess		
geta	Japanese wooden sandals	shoji	wall of sliding wooden panels
gohan	cooked rice		
haniwa	terracotta statuette	sukiyaki	Japanese dish
harakiri	(to commit-) see seppuku	sumo	wrestling, a sport much esteemed in Japan
hashi	bridge, chopsticks		
higashi, to	East	sashimi	Japanese dish
Honden	main hall of a Shinto shrine	sushi	Japanese dish
		sutra	Buddhist prayers
Hondo, kondo	main hall of a Buddhist temple	tatami	mat made of plaited rice straw (1 metre 80 by 90 centimetres)
ikebana	floral arrangement		
ji, in, do	temple	tempura	Japanese dish
jima, shima	island	tô	East, island, pagoda, lantern
jinja, gu, sha,	Shinto shrine		
Jigoku	hell	to	metropolis, capital
jo	castle	tokonoma	alcove in principal room of a house, contains a kakemono or an ikebana
kai	sea		
kaikyo	strait		
Kabuki	popular theatre	torii	Shinto gateway
kagura	Shinto dance	yukata	cotton kimono
kakemono	painted screen (vertical)	za	theatre building
kami	Shinto gods, spirits	Zen	Buddhist sect
Kannon	Goddess of Mercy		
kani	crab		
ken	prefecture, country		
kawa	river		
ko	lake		
kocha	tea		
koen	garden		
kondo	main hall of a Buddhist temple		
ku	district in a city		
machi	city		
makimono, emakimono	painted screen (horizontal)		
matsuri	festival		
minami	South		
minato	harbour		
Nishi, sai	west		
Noh	kind of theatre		
O cha	green tea		
onsen	hot spring		
patchinko	game, a sort of pin-ball or "flipper"		
ryokan	traditional Japanese inn		
sake	rice wine		
samurai	warrior, vassal of a daimio		
sha	Shinto shrine		
seppuku	suicide by the sword or dagger		
shinkansen	super express train, "bullet train"		
shima	island		
Shintoism	religion		
shogun	originally "sei-i-taisho-gun" = general in charge of defence against the barbarians, who eventually became a sort		

index

Abashiri (town) p. 88
Akan, p. 89
Akan-koan, see Akan
Akasaka, district, see Tokyo
Akiu onsen, spa, see Sendai
Amano Hashidate, p. 92
Ao shima, island, see Miyazaki
Arima onsen, spa, see Kobe
Arimura, belvedere, see Kagoshima
Asakusa, district, see Tokyo
Asakusa, temple, see Tokyo
Ashi, lake, see Fuji
Aso, p. 92
Atagawa onsen, spa, see Fuji
Atami, seaside resort, see Fuji
Atsuta Jingu, shrine, see Nagoya
Awaji-shima, island, see Kobe and Naruto

■
Babasaki bori, moats, walks, see Tokyo
Benten, caves, see Kamakura
Beppu, p. 93
Bihoro, see Akan
Bihoro toge, pass, see Akan
Bluff, district, see Yokahama
Bridgestone Art Gallery, see Tokyo
Bund, district, see Yokahama
Bunkyo ku, district, see Tokyo
Byodo-in, temple, see Kyoto

■
Castle of the Eagret, see Himeji
Castle of the Crow, see Okayama
Chino-ike, springs, see Beppu
Chion-in, monastery, see Kyoto
Chiyoda-ku, district, see Tokyo
Chubu, region, see Honshu
Chugoku, region, see Honshu
Chugu-ji, temple, see Nara
Chugushi, village, see Nikko
Chuo-dori, boulevard, see Tokyo
Chuo-ku, district, see Tokyo
Chuzenji-ko, lake, see Nikko
Confucian Shrine, see Nagasaki

■
Daibutsu, Great Buddha, see Kamakura
Daigyo, shrine, see Matsushima
Daiseki-ji, temple, see Fuji
Daisetsuzan, National Park, see Hokkaido
Daitoku-ji, monastery, see Kyoto
Daiya gawa, river, see Nikko
Dazaifu Temmangu, shrine, see Fukuoka
Dejima, island, see Nagasaki
Diet Building, see Tokyo

■
Engaku-ji, temple, see Kamakura
Enoshima, island, see Kamakura

■
Falls, Dragon's Head, see Nikko
Fuji-Hakone-Izu, p. 96
Fujinomiya, resort, see Fuji

Fuji san, Mount, see Fuji
Fukuoka, p. 99
Furansu dera, Catholic Church, see Nagasaki
Furasato, spa, see Kagoshima
Fushima Inari Jinja, shrine, see Kyoto
Futami, seaside resort, see Iseshima

■
Gegu Jingu, shrine, see Iseshima
Genkai, coastal park, see Fukuoka
Ginza, district, see Tokyo
Gifu, p. 100
Ginkaku-ji or Silver Pavilion, see Kyoto
Gion, district, see Kyoto
Glover House, see Nagasaki
Goryokaku, fort, see Hakodate
Gosho or Imperial Palace, see Kyoto
Gotemba, holiday resort, see Fuji
Gyokusen ji, temple, see Fuji

■
Hachiga minzoku bijutsu kan, museum, see Takayama
Hachiman gu, temple, see Takayama
Hakata, see Fukuoka
Hakodate, p. 102
Hakodate yama, volcano, see Hakodate
Hakone, see Fuji-Hakone-Izu,
Hakone Jinja, temple, see F.H.I.
Heian, shrine, see Kyoto
Heiwadai, park, see Miyazaki
Hibiya bori, see Tokyo
Hibiya, see Tokyo
Hida minzoku mura, museum, see Takayama
Hie, shrine, see Tokyo
Hiei, Mount, see Kyoto
Higashi gyoen, garden, see Tokyo
Higashi honga-ji, temple, see Kyoto
Himeji, p. 102
Himeji-jo, castle, see Himeji
Hirose gawa, river, see Sendai
Hiroshima, p. 104
Hokkaido, island of, p. 106
Honshu, island of, p. 109
Homotsu den, museum, see Tokyo
Hongaku-ji, temple, see Kamakura
Horyu-ji, temple, see Nara

■
Ichino machi, street, see Takayama
Ichinomiya, seaside resort, see Amano Hashidate
Idemitsu Art Gallery, see Tokyo
Ijuin, craft centre, see Kagoshima
Ikuta jinja, shrine, see Kobe
Inasa yama Hill, see Nagasaki
Io san, Mount, see Akan
Iro, Cape, see Fuji
Iseshima, National Park, p. 111
Ishibashi, bridge, Tokyo
Iso koen, Park, see Kagoshima
Isuju gawa, river, see Iseshima

INDEX 237

Itsukushima = Miyajima, p. 145
Itsukushima, shrine, see Miyajima
Izu, see Fuji-Hakone-Izu
■
Jufuku-ji temple, see Kamakura
■
Kabuki Theatre, see Tokyo
Kagoshima, p. 112
Kamaga take, Mount, see Fuji
Kamakura, p. 113
Kamakura ko kuho kan, museum, see Kamakura
Kano gawa, river, see Kyoto
Kamui Nupuri, lake, see Akan
Kanazawa, p. 117
Kanda, district, see Tokyo
Kansai or Kinki, region, see Honshu
Kanto, region, see Honshu
Kasamatsu, belvedere, see Amano Hashidate
Kashii gu, shrine, see Fukuoka
Kasuga Taisha, shrine, see Nara
Katsura, Imperial Villa, see Kyoto
Kawagughi ko, lake, see Fuji
Kegon Falls, see Nikko
Kencho-ji, temple, see Kamakura
Kenroku-en, garden, see Kanazawa
Kinkaku-ji or Golden Pavilion, see Kyoto
Kinkazan Island, see Matsushima
Kinka zan, hill, see Gifu
Kinki or Kansai region, see Honshu
Kyomizu dera, temple, see Kyoto
Kyosumi gardens, see Tokyo
Kobe, p. 117
Kobe o bashi, bridge, see Kobe
Kodai ji, temple, see Nagasaki
Kofuku-ji, temple, see Nara
Kofuku-ji, temple, see Nagasaki
Kokubun-ji, temple, see Takayama
Komachi koji, street, see Kamakura
Komitake, see Fuji
Kompira san, shrine, see Kotohira
Kongobu-ji, temple, see Koyasan
Koraku-en, garden, see Tokyo
Koraku-en, garden, see Okayama
Kotohira = Kompirasan Shrine, p. 119
Koya san, monastery, p. 120
Koyodai, see Fuji
Kyodo kan, museum, see Takayama
Kurakabe kan, museum, see Takayama
Kurashiki, p. 121
Kushiro, see Akan
Kutcharo ko, lake, see Akan
Kumamoto, see Aso
Kuttara ko, lake, see Noboribetsu
Kyoto, p. 124
Kyushu, island of, p. 141
■
Marunouchi district, see Tokyo
Mashu ko, lake, see Akan
Matsushima, p. 144
Matsushima kaigan, town, see Matsushima

Matsuyama, city, see Shikoku
Megane bashi, Spectacles Bridge, see Nagasaki
Meiji Shrine, see Tokyo
Meiji Shrine Inner Garden, see Tokyo
Meiji Shrine Outer Garden, see Tokyo
Memorial to the 26 Martyrs, see Nagasaki
Miya gawa, river, see Takayama
Miyajima = Itsukushima, p. 145
Miyanoshita, spa, see Fuji
Mihara, volcano, see Fuji
Miyazaki, p. 147
Miyazaki Jingu, temple, see Miyazaki
Monju, seaside resort, see Amano Hashidate
Motosu, lake, see Fuji
Moiwa yama, hill, see Sapporo
Museums, see list on next page
Myhon-ji, temple, see Kamakura
■
Nagara gawa, river, see Gifu
Nagasaki, p. 148
Nagoya, p. 152
Naigu Jingu, shrine, see Iseshima
Naka dake, crater, see Aso
Nakajima gawa, river, see Nagasaki
Nambam Museum, see Kobe
Nanzai san, mountain, see Nikko
Nara, p. 153
Naruto, p. 157
Narusawa, see Fuji
Nichinan, coastal park, see Miyazaki
Nijo, castle, see Kyoto
Nihombashi Bridge, see Tokyo
Nijubashi, harbour, see Tokyo
Nikko, p. 157
Nino Daira, open-air museum, see Fuji
Nino machi, street, see Takayama
Nishi Hongan ji, temple, see Kyoto
Nittai ji, temple, see Nagoya
Noboribetsu, p. 159
■
Odori, boulevard, see Sapporo
Oji machi, street, see Takayama
Okayama, p. 160
Oniyeama jigoku, springs, see Beppu
Osaka, p. 160
Osaki Hachiman gu, shrine, see Sendai
Oshima Island, see Fuji-Hakone Izu
Oshima, island, see Matsushima
Oyodo gawa, river, see Miyazaki
■
Palace, Akasaka Detached, see Tokyo
Palace, Imperial or Gosho, see Kyoto
Palace, Imperial, see Tokyo
Palace, Togu, see Tokyo
Park, Peace Memorial, see Hiroshima
Park of Peace, see Nagasaki
Pavilion, Golden or Kinkaku, see Kyoto
Pavilion, Silver or Ginkaku, see Kyoto
Panke, lake, see Akan
Penke, lake, see Akan
Pontocho, district, see Kyoto

Rinno-ji, temple, see Nikko
Rinno-ji, temple, see Sendai
Ritsurin park, gardens, see Takamatsu
Rokko san, hill, see Kobe
Roppongi, district, see Tokyo
Ryoan-ji, temple and stone garden, see Kyoto

■
Saboten, gardens, see Miyazaki
Sakunami onsen, spa, see Sendai
Sakurajima, volcano, see Kagoshima
Sagami Bay, see Fuji
Sanjusangen-do, temple, see Kyoto
Sanjo jinja, shrine, see Aso
Sankei-en, park, see Yokohama
Sapporo, p. 164
Saruga Bay, see Fuji
Sendai, p. 165
Sengaku-ji, temple, see Tokyo
Shiba Park, see Tokyo
Shibuya, district, see Tokyo
Shihorei, mountain, see Noboribetsu
Shikotsu-toya, park, see Sapporo
Shinjuku, district and park, see Tokyo
Shimbashi, district, see Tokyo
Shinobazu, lake, see Tokyo
Shikoku, island of, p. 166
Shimoda, fishing port, see Fuji
Shimotsui, fishing port, see Hokkaido
Shiraito no taki, waterfall, see Fuji
Shiraoi, Ainu village, see Noboribetsu
Shiroyama, hill, see Kagoshima
Shiroyama, hill, see Takayama
Shitamachi, district, see Tokyo
Shitenno-ji, temple, see Osaka
Sofuku-ji, temple, see Nagasaki
Sono machi, street, see Takayama
Shugaku in Rikyu, Imperial villa, see Kyoto
Sugimoto dera, temple, see Kamakura
Sumiyoshi, shrine, see Osaka
Susukino, district, see Osaka
Sumida, river, see Tokyo
Suwa-jinja, shrine, see Nagasaki
Suzaki, seaside resort, see Fuji-Hakone-Ize

■
Takamatsu, p. 166
Takasaki yama, mountain, see Beppu
Takatsuka, volcano, see Aso
Takayama, p. 168
Takayama jinya ato, museum at Takayama
Tatsumaki, springs, see Beppu
Tenjo san, mountain, see Fuji
Tenno-ji, park and temple, see Osaka
Terukuni jinja, temple, see Kagoshima
Toba, island, see Iseshima
Todai-ji, temple, see Nara
Tohoku, region, see Honshu
Tokyo, p. 170
Tomiyama, hill, see Matsushima
Tonosawa onsen, spa, see Fuji
Toshodai-ji, temple, see Nara
Tosho gu, temple, see Nara

Tsugaru, straits, see Hakodate and Hokkaido
Tsurumaru, castle, see Kagoshima
Tsurumi, springs, see Beppu
Tsurumi dake, mountain, see Aso
Tsurugaoka Hachinmaugu, temple, see Kamakura
Tsu sujigaoka, park, see Sendai
Tower, Tokyo the, see Tokyo

■
Udo jinja, shrine, see Miyazaki
Ueno Park, see Tokyo
Umi jigoku, spring, see Beppu
Usuki, Buddhas of, see Beppu

■
Wakakusa yama, hill, see Nara

■
Yakushi-ji, temple, see Nikko
Yakushi-ji, temple, see Nara
Yamanakako, lake, see Fuji
Yamanote, district, see Tokyo
Yamate, district, see Yokahama
Yashima, plateau, see Takamatsu
Yodogawa, river, see Osaka
Yokahama, p. 192
Yufu dake, mountain, see Aso
Yufuin onsen, spa, see Aso
Yunohira, belvedere, see Kagoshima
Yumoto onsen, spa, see Fuji
Yunokawa onsen, spa, see Hakodate

■
Zuigan-ji, shrine, see Matsushima
Zojo-ji, temple, see Tokyo
Zozu san, hill, see Kotohira

■
Museum, Ainu, see Sapporo
Museum of Archaeology, see Kurashiki
Museum, Electrical and Science, see Osaka
Museum of Folkcraft, see Kurashiki
Museum, Fujita, see Osaka
Museum, Marine and Harbour, see Kobe
Museum, National, see Kyoto
Museum, National, see Nara
Museum, National, see Tokyo
Museum, National Science, see Tokyo
Museum of Modern Art, National, see Tokyo
Museum, Municipal Fine Art, see Osaka
Museum, Nezu, see Tokyo
Museum, O Hara Magosaburo, see Kurashiki
Museum, Okura, see Tokyo
Museum, Silk, see Tokyo
Museum, Sumo, see Tokyo
Museum, Theatre, see Tokyo
Museum, Toy, see Kurashiki
Museum, Tokugawa, see Nagoya
Museum of Western Art, National, see Tokyo

japan
today

series editor   jean hureau
photographs by   jacques marthelot
                     except when otherwise credited

© 1978
éditions j.a.
51, avenue des ternes - 75017 paris
all rights reserved

printed in france
printing completed 3rd quarter 1978
legal copy deposited 3rd quarter 1978
publisher's n° 1202/1
ISBN 2-85258-110-8

in the same series

by jean hureau	■ iran today *2nd edition*
	■ egypt today
	■ syria today
	■ corsica today
	■ tunisia today
	■ provence and the french riviera today
by raymond morineau	■ lebanon today
by mylène rémy	■ senegal today
	■ ivory coast today
	■ ghana today
	■ gabon today
by jacques-louis delpal	■ paris today
	■ the valley of the loire today
by jacques legros	■ scandinavia today
by siradiou diallo	■ zaïre today
by louis doucet	■ the caribbean today
by anne debel	■ cameroon today
by george oor	■ yugoslavia today
by maurice piraux	■ togo today
by clarisse desiles	■ japan today
in preparation	■ morocco today
	■ great britain today